CHESAPEAKE WATERS

CHESAPEAKE WATERS

Four Centuries of Controversy, Concern, and Legislation

SECOND EDITION

To Marcella
Thank you
for your
interest in
Chesapeake Bay

Steven Davison

BY
STEVEN G. DAVISON
JAY G. MERWIN, JR.
JOHN CAPPER
GARRETT POWER
FRANK R. SHIVERS, JR.

Tidewater Publishers
Centreville, Maryland

Library of Congress Cataloging-in-Publication Data

Chesapeake waters : four centuries of controversy, concern, and
  legislation / by Steven G. Davison . . . [et al.]. — 2nd ed.
     p.   cm.
  Previous ed. cataloged under: Capper, John, 1937– .
  Includes bibliographical references and index.
  ISBN 0-87033-501-4 (hardcover)
    1. Water—Pollution—Health aspects—Chesapeake Bay Region (Md.
  and Va.)—History.   2. Marine pollution—Health aspects—Chesapeake
  Bay (Md. and Va.)—History.   3. Water—Pollution—Environmental
  aspects—Chesapeake Bay Region (Md. and Va.)—History.   4. Marine
  pollution—Environmental aspects—Chesapeake Bay (Md. and Va.)—
  History.   5. Water—Pollution—Chesapeake Bay Region (Md. and
  Va.)—Public opinion—History.   6. Marine pollution—Chesapeake Bay
  (Md. and Va.)—Public opinion—History.   7. Environmental policy—
  Chesapeake Bay Region (Md. and Va.)—History.   8. Chesapeake Bay
  Region (Md. and Va.)—History.   I. Davison, Steven G. (Steven
  Gebauer)
  RA592.C45C44   1997
  363.739′4′0916347—dc21   97-28122

Manufactured in the United States of America
First edition, 1983. Second edition, 1997

For John Capper (1937–92)

Contents

Preface to the Second Edition

In this new edition the authors bring up to 1996 their account of how public opinion and politics have affected Chesapeake waters for nearly four hundred years. Their first edition surveyed history from 1607 to 1972, the date of the destructive Tropical Storm Agnes. To bring the record up to date, Steven Davison and Jay Merwin came on board. They and the original authors decided to retain most of the original text and illustrations. In the earlier edition Frank Shivers covered the first 250 years of settlement and found illustrations. John Capper wrote of the complex modern period with help from Garrett Power, who coordinated work on both editions. Now this second edition updates the first by surveying the substantial amount of new information about the Bay. It also examines governmental programs established during those twenty-four years to protect and restore the Chesapeake.

Almost simultaneously with the publication of the first edition of this book in 1983, the U.S. Environmental Protection Agency (EPA) published its massive study of the Chesapeake Bay. EPA's study concluded that the Bay's water quality and aquatic life had seriously declined (particularly as a result of the runoff of phosphorus and nitrogen [nutrients] from farms in the Bay's watershed, which cause algae blooms in the Bay that result in less oxygen and fewer submerged grasses, fish, and shellfish in the Bay). The EPA study also recommended that the federal government and the states within the Bay's watershed join together in an interjurisdictional program to protect and restore the Bay.

In response to EPA's findings and recommendations, the federal government, Maryland, Virginia, Pennsylvania, and the District of Columbia in 1983 launched the Chesapeake Bay Program, a blueprint for managing Bay restoration and protection programs on an interjurisdictional basis. In the following year, Maryland enacted the Chesapeake Bay Critical Area Protection Act to minimize runoff pollution from development and agricultural and forestry activities on lands near the Bay and its tributaries. Virginia enacted its own legislation to protect the Bay from development in 1988.

In the thirteen-plus years since release of EPA's Bay study, scientific knowledge about the causes of the Bay's decline has increased substantially. And during this period the Bay program has improved its water quality and ecosystems in many respects.

The Chesapeake, however, still is being significantly harmed by the runoff of nutrients after tropical storms, heavy rains, and snow melt-off. Furthermore, uncontrolled emissions of nitrogen into the air from automobiles, power plants, and industrial facilities also are causing significant harm to the Bay.

Generally, however, we have learned in the last fifteen or so years that the Bay's problems are the result of the cumulative effects of the modern lifestyle and economy, from driving automobiles to developing land.

The revised and new chapters in the second edition of this book provide an overview of the knowledge that we have gained about the Bay in the last twenty-five years and of the actions that the federal government, Maryland, Virginia, Pennsylvania, and the District of Columbia have taken during this period to restore and protect the Bay. In this second edition, we examine how far we have come in efforts to return the Bay to health and the difficult economic and political choices that lie ahead in furthering its recovery.

The authors thank the Maryland Chesapeake Bay Critical Area Commission and the Alliance for the Chesapeake Bay for allowing us access to their libraries and slide collections and for allowing us to use slides from their collections for illustrations for the second edition. The authors also are grateful to Indiana University Press for permission to quote from Abel Wolman's *Water, Health and Society* (1969); to Johns Hopkins University Press for permission to quote from Susan Q. Stranahan's *Susquehanna, River of Dreams* (1993); and to the Virginia Museum of Natural History Foundation for permission to quote from Gerald L. Baliles's *Preserving the Chesapeake Bay* (1995). Special thanks are extended to Shari Wilson, of the Alliance for Chesapeake Bay and the Maryland Department of the Environment, and to Ren Serey, executive director of the Maryland Chesapeake Bay Critical Area Commission, for reviewing the manuscript. We also are very grateful to Dawn Taylor and Donna Pennepacker for their assistance in preparing the manuscript.

Preface to the First Edition

The Chesapeake Bay is the most studied and best understood estuary in the United States. Yet, it is practically unexamined in the areas of the social sciences and the humanities. While millions of dollars have been spent on producing the thousands of studies that examine the physical, biological, chemical, and engineering aspects of the Bay, little attention has been given to understanding the political, cultural, and economic character of Bay governance.

The relationship of the governments of Maryland and Virginia to the Bay is imperfectly documented. Government documents which do exist are scattered in various libraries in both states and have not found their way into the numerous bibliographies that have been assembled for the Bay. In Virginia, the State Water Control Board did not produce annual reports until 1972, the cutoff date for this study. In Maryland, the reports of water-quality agencies tend to be perfunctory and repetitive, and they give little indication of the real issues facing the agencies over the years. The many planning documents which do exist (the recent Corps of Engineers' Chesapeake Bay Study is the largest) are general compilations of information and issues rather than original pieces of research.

As a result, the present study has had the benefit of little scholarship to point the way. The researcher is forced to approach his material as though he were an archeologist, finding a few shards here, a few bone fragments there. Piecing together a coherent story out of the fragments requires a certain amount of logic, a workable hypothesis about the overall nature of the creature to be described, and some theories about how the evidence fits together.

But the story is worth the telling. After all, the quality of Chesapeake Bay is a matter of public opinion as well as scientific opinion. Those concerned about the Bay must understand the human-political dimension as well as the physical-biological side.

We relied primarily on written sources. Those proving most fruitful have been the annual reports of various state agencies, the occasional reports of study commissions and blue ribbon panels, and the codes, statutes, and case law of the two states. Agency files proved difficult to use because they are boxed and stored, full of irrelevant material, unorganized, and uncataloged.

Interviews with persons familiar with Bay issues have given a general orientation to a particular period and suggestions of topics of sources for further research. We have not attempted to get detailed information of specific events through such interviews. The written record, we feel, stands on its own.

In particular, we also made use of the abundant collections of newspaper files in libraries. While newspaper articles may have questionable accuracy, they identify key issues and place them definitively in time. Without them, numerous controversies, left only to the official archivists, would go unrecorded. In this study, information from newspapers gives a sample of issues and shows the broad trends in water-quality awareness. Feature articles in magazines and newspapers are particularly useful, because they both reflect, and partially shape, the public attitudes toward the Bay. Changes in these attitudes provide data used throughout the report.

We hope that this book has something to say that has been neglected in the public debate over the Bay. Its conclusions do not mean that scientists should be involved less in research on the Bay. They simply suggest that economists, political scientists, historians, and lawyers, should be involved more.

The authors thank Randall Beirne, William A. Cook, Marian Dillon, Richard I. McLean [presently an administrator with the Maryland Department of Natural Resources], David P. Miller, Sandralee P. Morris, Erik Olson, Natalie W. Shivers, LuAnn Young, H. Chace Davis, Jr., and Scott Lewis Zeger, all of whom made special contributions. We are likewise indebted to Penelope R. Power who prepared the index. The authors are also grateful to Gilbert Byron and his publisher, The Driftwind Press [and to Chesapeake College], for permission to quote from "Tangier Prayer" (1942), to Annie Dillard and her publisher, Harper's Magazine Press, for permission to quote from *Pilgrim at Tinker Creek* (1974), and to Alfred A. Knopf, Inc. for permission to quote H. L. Mencken from *Newspaper Days* (1941) and *Happy Days* (1940).

Preparation of the initial report upon which this book is based was funded in part by the Chesapeake Bay Program of the U.S. Environmental Protection Agency (EPA Grant X-003226-01). It has been approved for publication. Approval does not signify that the contents necessarily reflect the views and policies of the U.S. Environmental Protection Agency, nor does mention of trade names or commercial products constitute endorsement or recommendation for use.

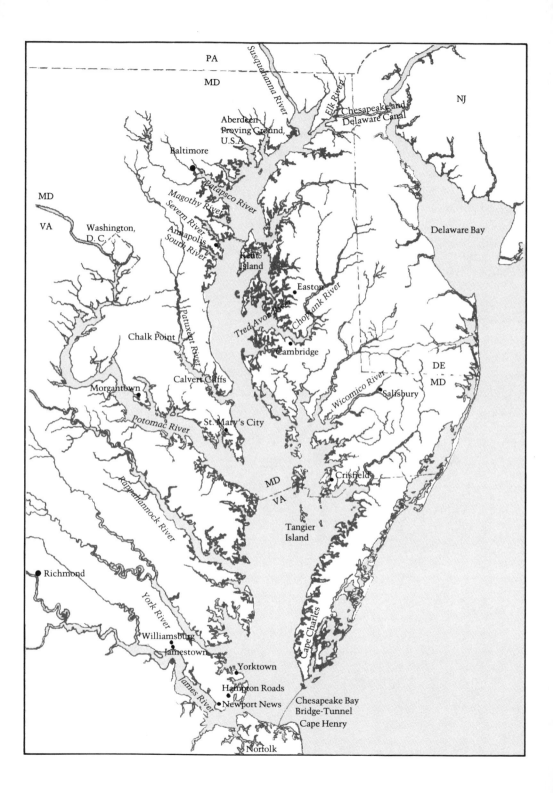

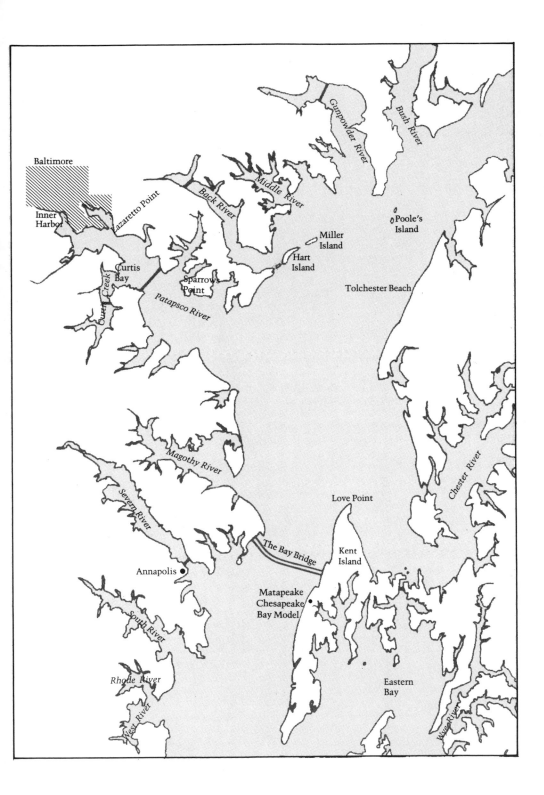

CHESAPEAKE WATERS

1

From Jamestown to Contract with America

---

*There were never Englishmen left in a Foreigne Country in such
miserie as we were in this new discovered Virginia. . . . Our drinke
cold water taken out of the River, which was at a floud verie salt, at a
low tide full of slime and filth, which was the destruction of many of
our men.*

—George Percy, 1625

*...The severely distressed condition of the Bay has been well
documented. Abuse, neglect and lack of adequate care are the causes.
In short, we've been overloading the Bay with nutrients, sediments
and dangerous chemicals, poisoning the water in the Bay bottom with
toxic pollutants, and destroying wetlands and grasses which are key
ingredients necessary for a healthy and productive Bay.*

—Former Virginia Governor Gerald L. Baliles, 1995

In 1607, the first English colonists in the Chesapeake Bay region
drew their drinking water from the often brackish James River estuary, which ebbed and flowed past the shores of the swampy island
that was to be called Jamestown. In doing so, they exposed themselves to disease-causing bacteria from their own wastes (according
to the theory of a twentieth-century historian) and subjected themselves to gradual salt poisoning from the brackish water.[1]

These subtle and speculative conditions of 1607 and the pollution of the Bay from the effects of modern-day living bracket this
book, a historical survey of water-quality issues on the Chesapeake
Bay from 1607 and European settlement to 1996 and modern mega-
lopolis, from the microscopic and localized to the obvious and Baywide. This survey illustrates that water quality, although affected by
human activity, is to a large degree independent of what man does
and that sophisticated tools of science are often required to detect
who is having what effect on the water.

Landing at Jamestown. Engraved for the *New York Mirror,* May 8, 1841. This nineteenth-century romanticized version of the first permanent settlement by Europeans north of Spanish America is reminiscent of the exaggerated idealization of settlement prospects reflected in much of the writing of the early seventeenth century about the New World. Courtesy the Library of Congress.

Since 1607, the people of the region have taken a great interest in the Chesapeake Bay, on account of its vast expanse and its productivity as a fishery, the latter making it commercially important to people worldwide. The Bay also has long served as a vital artery of transportation as well as the scene of various forms of recreation. And it has always been home to a rich and unusual mix of plants and wildlife. As Governor Baliles puts it: "[W]e as humans cannot measure *this* value of the Bay. This is just one root of the watershed's current problems. With so many dependent upon the Bay for their incomes, or their lives, conflict over the best use of the Chesapeake's resources has inevitably developed. Choices must be made."[2]

This book traces the development and evolution of governments' attention to the Bay. The focus is on human institutions, ideas, and attitudes, as well as on the Bay's physical, chemical, and biological attributes. We examine how ideas about water quality have been formed, changed, and communicated; how government has organized itself to the task of studying and influencing the quality of the Bay; and how perceptions of the Bay have brought about change in the way it is used.

Our thesis comes from the view that, while it is important to describe the conditions of the Bay as accurately as human science and technology allow, it is just as important to understand how the concept of water quality has operated in the political process. This book, then, is an attempt to broaden the debate about the Bay and its future and to explain how we and our predecessors have used our governments to influence the quality of Bay water.

Now for the definition of a few terms: *Government* is broadly interpreted to include any activity or viewpoint sanctioned by government bodies. It includes actions by courts, legislatures, and administrative agencies—local, state, or national. During the nearly four hundred years of recorded history of the Bay, there have been shifts in the locus of governmental power. In the seventeenth and eighteenth centuries, the courts were the dominant law-making bodies. By resolving a number of ad hoc conflicts, they created what came to be viewed as a body of common law. Then, during the nineteenth century, the Maryland and Virginia legislatures attempted to address some problems of water quality. They enacted laws designed to resolve specific conflicts over fisheries and navigation.

Early in the twentieth century, talk grew of managing water quality by administrative agencies, although the word *managing* may be a misnomer. Agencies do not manage the Bay in any generally accepted sense of the word. Although human influence on the Bay (both intentional and unintentional) has been considerable, the Bay is still to a large extent beyond the ability of our technology to regulate or control. Nonetheless, an "administrative state" has evolved in which public administrators influence the Bay to a greater extent than courts or legislatures.

Our definition of the *Chesapeake Bay* is expansive. In a physical sense it includes the whole estuary—the semienclosed coastal body of water that has a free connection with the open sea and within which seawater is measurably diluted by fresh water derived from land drainage. So defined, the Bay includes long stretches of the Potomac, Rappahannock, York, and James rivers, as well as hundreds of other rivers, creeks, bays, and sounds. The nomenclature of the Bay region is confusing. Some rivers are not fresh and flowing but subestuaries (for example, the Severn River); other rivers flow to the fall line and then become estuarine (for example, the Potomac River). Likewise, some creeks are tidal and others not. The words *bay* and *sound* are also loosely applied. Pocomoke Sound fits the conventional definition of a bay while Chincoteague Bay behind the barrier beach on the Atlantic is perhaps more appropriately a sound. This study avoids semantic confusion by looking to the estuary regardless of the popular name.[3]

This study also looks beyond the estuary. Because the quality of the Bay is the result of all physical additions, we consider activities in the entire Chesapeake Bay watershed, which includes six states and reaches as far from the Bay as mid-New York State. The six states are Delaware, Maryland, New York, Pennsylvania, Virginia, and West Virginia; to these, add the District of Columbia. The chief boundary dispute around the Bay—between Maryland and Virginia—still causes trouble among watermen.

A historical perspective does give meaning to the concept of water quality in the Chesapeake Bay. For roughly the first two hundred years covered by this book, the only water-quality issues were siltation, the condition of the fisheries, and public health. By the 1820s concerns were growing about the declining stocks of shad and herring. These declines were attributed in part to the effects of siltation, sedimentation, temperature changes due to land clearing, and the mechanical action of ships' wakes, as well as the blockage of streams by dams and raceways.

By the middle of the nineteenth century, people also recognized that fish avoided areas that were fouled by wastes, such as sawdust, canning wastes, and discharges from slaughterhouses and tanneries. The present usage of the word "pollution" then came about. To-

Two pencil sketches by John H. B. Latrobe illustrate uses of the upper Chesapeake about 1830. *Top,* commercial ships at the mouth of the Susquehanna River, Turkey Point, Maryland. *Bottom,* a raft on the Elk River, Maryland, on its way to a Chesapeake port. These calm waterscapes contrast with the flood of fresh water caused by Tropical Storm Agnes (1972), which so damaged the Bay. The Susquehanna River, even when not in flood, is the Bay's greatest single source of fresh water. Courtesy the Library of Congress.

ward the end of the century, fisheries scientists were making the first formal studies of the relation of fish life cycles and productivity to environmental factors, both natural and man caused. By the twentieth century, the germ theory of diseases had added the concept of waterborne disease bacteria to the growing science of Bay water quality.

Once it was established that oysters were carriers of diseases originating from human sewage, oyster sanitation became the most influential water-quality concern. This connection was made in the 1890s, but had its first impact on the Bay shortly after 1900. Also about this time fishermen began complaining about the effects of wastes from food-processing plants. Thus began the concern for industrial wastes that for the first half of the century continued to come primarily from food-processing industries and other small producers of organic wastes.

By the end of World War I, oil from ships using the Bay was seen by fisheries officials and bathing beach operators as a major water-quality problem, and in the mid-1920s, oyster sanitation again became a major concern when several epidemics blamed on bad oysters resulted in a widespread closure of bars. By the end of World War II, growth of municipal sewerage and industry created conditions that made pollution in both tidal and nontidal streams a major concern.

In the 1950s, scientists began to investigate more subtle water-quality factors, such as the effects of insecticides, fertilizers, and heavy metals on marine organisms. That investigation ushered in the era of scrutiny of exotic discharges to the system, which remain a predominant concern today. By about 1960, a new set of issues was added to the list, including thermal discharges, overenrichment, general land runoff, and freshwater diversions. Also in the 1960s, attention expanded from a focus on specific conditions and pollutants to a general worry about the overall ecological health of the Bay: that is, the context in which water quality was viewed shifted from pollution to ecology.

During the 1970s, as the environmental movement quickened with the first Earth Day, the federal government funded the construction of improved sewage treatment plants along the Bay and its tributaries, leading to reduction in discharges of the harmful nutrients phosphorus and nitrogen. The pollution of the James River by the pesticide Kepone received national attention in the mid-1970s. The EPA began a massive study of the Bay and declared it to be in a state of serious decline, owing largely to runoff of nutrients from farms and to the changes in land use of previously forested land. Confronted with the spoiling of such a large body of water crossing many state and local jurisdictional lines, the states around the Bay joined with the federal government in a regional approach that took the form of the Chesapeake Bay Agreements of 1983, 1987, and 1992. From those accords sprang the Chesapeake Bay Program, a voluntary undertaking involving federal, state, and local governments and private individuals and businesses aimed at reversing the Bay's decline. But despite some success, the Bay remains a system of troubled waters; restoration and protection of the Bay will require the expenditure of hundreds of millions of dollars by government and the private sector for years to come.

This study takes a historical look at the quality of Chesapeake Bay waters and the measures that governments have taken to influence that quality. Because most conflicts about the Bay involve the issue of water quality, this book in some respects is a history of Bay uses in general. The organizing principle and point of view, however, stress the relation of various uses to water quality and the ways government has intervened to have some effect on that quality.

The book's organization is both chronological and thematic. We describe the nature of the Bay in chapter 2, and the uses and abuses of Chesapeake waters during the first three hundred years of European settlement in chapters 3, 4, 5, and 6. In chapters 7, 8, 9, 10, and 11 we take up the major conflicts that have arisen with respect to the Bay in the twentieth century, in more or less chronological order. In chapter 12 we analyze the present condition of the Bay, and the types and sources of pollution and activities that have caused the decline of water quality and aquatic resources of the Bay in recent years. Chapter 13 analyzes recent actions that the Bay states and federal government have taken to restore and protect the Bay. In chapter 14, we state our conclusions.

2

Noble Arm of the Sea

The Chesapeake Bay is a noble arm of the sea.

—Lord Morpeth, 1842

Virtually all written documents about the Chesapeake Bay, whether technical or popular, begin with a description of the Bay and the region. Because these descriptions contribute to the composite public perception of the Bay, we would like to make some general observations about these Bay descriptions before presenting our own.

CREATION OF THE BAY

Until recently, scientists theorized that the Chesapeake Bay was formed as a result of ocean levels rising after glaciers retreated and melted at the end of the last Ice Age. A million years ago, glaciers had moved up and down the Chesapeake area, shaping the land beneath them. According to conventional theory, the level of ocean waters rose and the sea flowed over the continental shelf as glaciers melted at the end of the Ice Age. Thereafter, according to this theory, ocean waters continued to rise, reaching the present mouth of the Chesapeake Bay ten thousand years ago and then rising further until the valleys at the mouths of the Susquehanna, Potomac, Rappahannock, York, and James rivers were submerged. Scientists have theorized that these flooded river valleys formed today's Chesapeake Bay basin, with the Bay reaching its present dimensions approximately three thousand years ago.

Recently, however, a group of scientists, including C. Wylie Poag of the U.S. Geological Survey and Christian Koeberl of the University of Virginia, has presented evidence supporting a theory that the formation of the Bay is due at least in part to a huge meteor smashing into what we now call the lower Chesapeake Bay, slightly west of

Cape Charles, Virginia, at the tip of the Delmarva peninsula, some thirty-five million years ago. At the time, ocean levels were much higher than they are today, and the meteor crash site was covered by ocean waters as deep as 1,500 feet.

Other scientists are skeptical of the Poag-Koeberl meteor theory, noting that it is not yet supported by hard evidence such as the discovery of minerals displaying shock patterns caused only by a very powerful explosion or containing abnormal concentrations of the metallic element iridium, which is attributable to meteors from outer space.

The Poag-Koeberl group believes, however, that a meteor at least a mile in diameter crashed and exploded on impact in a huge fireball, incinerating plants and animals in ocean waters and near the shore. The entire earth would have been shaken, as if by a huge earthquake, and hot gases and molten rock would have spurted many miles into the earth's upper atmosphere. The collision may have caused glass (formed from crystallized molten rock) to rain down as far away as New Jersey, Texas, and Barbados, and a tsunami as high as 50 feet to hit the seacoast.

The Poag-Koeberl group has found evidence that this giant meteor created a crater 55 miles in diameter. Poag speculates that the rivers presently flowing into the Bay do so because their waters sought the low ground of the meteor's crater. Thus the blasting of the crater may have begun the formation of the Bay by funneling rivers into it, creating an indirect ". . . template for later development of the Bay."[1]

DESCRIBING THE BAY

In general, descriptions of the Bay show the lack of a single authoritative source of information about this body of water, its resources, and the region that surrounds it. Those descriptions that do exist are sometimes inaccurate and present problems of definition and interpretation. Several examples will illustrate these problems.

One might expect that physical dimensions of the Bay, which are for the most part constant and accessible to measurement, would be readily available and accurate. Such is not the case. Anyone examining the literature of the Bay, whether recent or old, technical or popular, will find wide discrepancies in the various dimensions.

Length, for example, is one such dimension. One would expect to find some variation because there is some judgment involved in such matters as placing the starting and ending points. Where does the Bay end and the ocean begin, at the mouth? And where does the Bay begin and the Susquehanna end, at the head? Another question is how to deal with the curves of the Bay. Nonetheless, these considerations do not account for the wide variation.

In fact, the length of the Bay has been authoritatively reported to be approximately 190 miles, using a method of measurement de-

Captain John Smith's 1627 map of Virginia, the chief source of information on the New World for sixty years. Courtesy the Library of Congress.

scribed in a Chesapeake Bay Institute special report.[2] We came very close to duplicating this figure using a 1:200,000-scale chart and a map wheel. Yet, in the Corps of Engineers' massive Chesapeake Bay study, the Bay is reported to be over 200 miles long,[3] and in 1979, two widely published Bay authorities coauthored a paper in which the length is listed as 168 miles (271 kilometers), which is even less than the straight-line distance between the lines of latitude of the mouth and head of the Bay (37 degrees north to 39 degrees 32 minutes, or 152 nautical miles, which equal 175 statute miles).[4] Hence, while there are sources of accurate information, it is difficult for all but the most diligent reader to sort out the good from the bad. Because many writers of descriptions draw on previously published sources without question of citation, old errors are perpetuated.

Problems of definition complicate efforts to relate census data to the Bay. Descriptions that give population figures, and at least imply that there is a direct relation between population and the stresses of pressures put on the Bay, pose two questions. First, there is always a question of how to define the Bay region. Descriptions might range from including a narrow strip of land around the Bay to including the entire watershed. Second, and more perplexing, is the question of the relation of human population to the Bay. Impact on the Bay differs, depending on location, method of sewage disposal, predominant

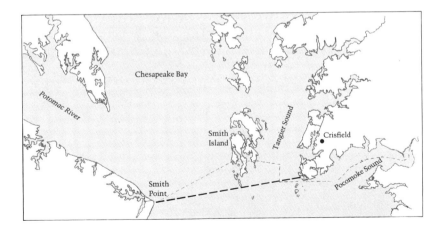

The section of Chesapeake Bay showing the shifting boundary between Maryland and Virginia, long a matter of conflict. The dark line is the boundary of 1668 and the light line is the present boundary.

land use, and employment. For example, five thousand watchmakers living in metropolitan apartments may have less impact on the Bay than one dairy farmer living on the shores of an oyster-producing creek. As a result population figures may then lack the analysis necessary to make them relevant to Bay issues. That said, the raw data of Bay *watershed* population are as follows: the area held 8.4 million residents in 1950 and 14.7 million residents in 1990. By the year 2020, the number is expected to swell to 17.4 million.

Perhaps no numbers signify more for Bay water-quality issues than fisheries statistics. Much of the political energy that drives the Bay system is generated over public concern for the state of the fisheries. Problems associated with the collection of these numbers are legion: Methods of collecting information have changed markedly over the years; methods vary between the states; figures for sport catches are unknown; and harvests are affected by economic and cultural forces having nothing to do with the size of the fish population (low prices and war, for example), and fishermen suspected of underreporting catch to avoid taxes. Moreover, fluctuations in fish populations and, therefore, catch, are subject to wide variations due to unknown or poorly understood factors. These factors could be natural or human in origin.

Serious study has gone into the examination and analysis of fisheries statistics and there is continuing advance in their utility as management tools.[5] But over the historical period covered by this book, these statistics have tended to be used uncritically, particularly when inferences are drawn from them about the effect of man's activities on the water quality of the Bay.

CHARACTERISTICS OF THE BAY

Forewarned by our own criticism, we now undertake our description of the Chesapeake Bay. It is intended to be relevant to the specific topic of study contained in this book: man's effect on water quality over time.

Geographic Characteristics

Common practice compares the Bay to other large estuaries or lakes. One of its striking features is that it has a very long shoreline in relation to its overall length or size. Taking the previously discussed length figure of 190 miles and a currently prevalent figure for shoreline of about 8,000 miles, the Bay has a shoreline-to-length ratio of more than forty to one. Lake Ontario, the smallest of the Great Lakes, has a ratio of about two and a half to one. For the Chesapeake, there are roughly 2 miles of shoreline for every square mile of surface area; for Ontario, there is about 1 mile of shoreline for 18 square miles of surface area.

These ratios point to at least three important aspects of the Bay. First, they are another way of expressing the extreme branchiness of the Bay: the seemingly numberless bays, creeks, and guts that so frustrated colonial land travelers, but so fascinate the contemporary "gunkholer." Second, the extensive shoreline allows for a high level of human access to the Bay, both to use and to enjoy, and thus to affect it by land activities. Third, the land-water edge is an area of relatively high biological activity, and probably contributes materially to the productivity and biological variety of the Bay.

Another striking characteristic is its shallowness. The Bay averages only approximately 27 feet in depth—21 feet if the depth of the Bay's tidal tributary rivers are included. An estimated 10 percent of the Bay has a depth of less than 3 feet and 20 percent of the Bay is estimated to have a depth of less than 6 feet, 6 inches.

This was apparent to the visitor of the Bay hydrologic model in Matapeake, Maryland. The model, housed in a massive building covering 14 acres, was closed in 1983 after ten years of use because advances in computer modeling and simulation made it obsolete. In the massive, still-standing building in Matapeake is a concrete Bay in miniature, at a horizontal scale of 1,000 to 1. The model itself is more than 1,500 feet long (equal to the length of five football fields). Since it is enclosed under a single roof, there is a sense of vastness about the place. Yet the greatest depth of the Bay model is only 18 inches. When it was in use, most of the underwater area was 3 inches or less deep. It must then be remembered that the vertical scale is only 100 to 1 (10 percent of the horizontal ratio). Thus, if the model were at the same scale in both directions, the deepest part would have been about 2 inches when the model was in use, and most of the "Bay" would have been covered by a film of water scarcely worthy of being called wet.

Again, a comparison to Lake Ontario is in order. There, a similar scale model would have a depth of 8 feet at its deepest, and a large underwater area over 5 feet deep. It would make a fine swimming pool, whereas the Bay model in most places wasn't even deep enough for a toddler to wade in. In terms of how much water it holds, the Bay proper, with its 18 trillion gallons, is a veritable drop in the oceanic bucket. Lake Ontario contains approximately forty times as much water. If dilution is the solution for pollution, the Chesapeake has a relatively limited capacity.

Finally, it is popular to speak of the Bay as a single system. Yet, in a profoundly important way, it is divisible into many separate sub-systems, in a manner that many other large bodies of water are not. The rate at which water is exchanged or flushed from a subestuary to the estuary is often small or negligible. One effect of this geographic feature has been to make local conditions more acute than they might otherwise have been at the point of discharge. For example, a sewage treatment plant in Baltimore has fundamentally altered the quality of Back River, on which it is located. On the other hand, effects from these wastes are more isolated and contained than they would have been if discharged into the main stem of the Bay.

Demographic Characteristics

We have already spoken of the difficulty of relating raw population figures in any direct way to the condition of the Bay. However, several potentially important generalizations are possible. First, municipal-industrial influences, whatever they might be, are concentrated in a relatively small area in and around each of the Bay's major cities of Baltimore, Washington, Richmond, and Norfolk. The remainder of the Bay region is relatively lightly populated. Second, these urban areas have all experienced rapid growth since World War II and will probably continue to do so. Land use has changed substantially within the standard metropolitan statistical areas, from forested and agricultural to residential and commercial, with attendant effects on stream runoff and waste loads.

Third, the region as a whole has substantially less heavy industry than comparable regions in the East or Midwest.[6] Employment in government and service sectors, including the military, is higher than the regional average. Fourth, even though the region has the image of being an important seafood producer, and fishing provides an important source of employment in some of the smaller tidewater coun-

ties of both states, from a regional standpoint the contribution of the fisheries industry to the economy is almost negligible.[7]

Physical-Chemical Characteristics

Along with demographic characteristics, we must understand the major physical and chemical features of the Bay. These have been succinctly and clearly stated by Dr. Donald Pritchard, a leading Bay scientist who was director of the Chesapeake Bay Institute.[8] Several important aspects might be usefully highlighted for the layman. First, waters in the Bay and its principal tributaries are in constant motion, gravity acting differentially on the denser salt water being overridden by the lighter fresh water entering from the rivers. This two-layered wedge effect accounts for the presence of seawater (much diluted) near the heads of the Bay and its tributaries, and the eventual exit of river water to the sea. This basic circulation is much affected by variations in river flow and combinations of winds and atmospheric pressure changes that can drive very large quantities of water into or out of the Bay.

Second, the circulation of many of the smaller arms of the Bay is relatively sluggish because they do not have a substantial freshwater inflow to establish their own flow system. Their circulation is much affected by the differences in salinities between their lower reaches and that of the Bay. At certain times of the year they can undergo a rapid flushing because of strong differences between their salinity and that of the Bay. At other times there may be very little exchange. In some respects, then, for substantial periods of the year, these smaller bodies of water are essentially separate from the Bay. This condition applies as well to the smaller branches of the principal tributaries.

Third, the fate of a constituent added to the waters of the Bay is often complex. Silt from land erosion may interact, for example, with bacteria from a farm feedlot; nutrients from a sewage treatment plant might combine chemically with the wastes from an industry. A toxic chemical might be buried harmlessly in bottom sediments, or it might be physically and biologically transported throughout the Bay.

Fourth, alterations made by man can have a considerable impact on the physical and chemical oceanography of the Bay. A system of dams might substantially alter the flow regimen of a river entering

the Bay. A navigation channel might affect the exchange rate between a tributary and the main Bay. Industrial and municipal wastes, moreover, can greatly increase the concentrations of naturally occurring substances or add new and potentially harmful ones.

On the other hand, the natural forces affecting the Bay, and their range of variation, are enormous when compared in general to the scale of effects that are influenced or controlled by man. It is important to keep these forces in perspective when reviewing the history of humankind's concern for its impact on the quality of the Bay.

Biological Characteristics

A fitting companion to Dr. Pritchard's paper just cited, and bound next to it in the report of the 1968 Governor's Conference on the Chesapeake Bay, is the description of the biology of the Bay by Dr. L. Eugene Cronin of the Chesapeake Biological Laboratory, another leading Bay scientist.[9] First, as the Bay is physically and chemically a highly variable environment, so are the biological populations that fill the various Bay niches. Populations and reproductive success of numerous important species, including the oyster, the crab, and the striped bass, are subject to cycles that are as yet only partly comprehended. Understanding of these natural variations is confounded by the problem of assessing the influence of humans, either through their harvests or through their indirect effect on water quality.

Second, an accounting of the animals of the Bay combines permanent residents, regular and Bay-dependent visitors (sometimes in great abundance), and strays. The biological openness of the system is one of its prominent features.

Third, the biology of the Bay poses a paradox. The Bay combines toughness and resiliency to change, on the one hand, and fragility and vulnerability, on the other—the former because the residents of the Bay are adapted to its great natural variability, and the latter because of the relatively short food chains in the Bay.

Although the goals of the 1987 Chesapeake Bay Agreement include the Bay's "entire natural system" being "healthy and productive," no one is quite sure what would constitute full health and productivity of the Chesapeake. When John Smith first explored the Bay in 1608, it probably contained large numbers of oysters, bottom-dwelling fish (such as sturgeon, sheepshead, red drum, black drum, and perch), and other bottom-dwelling species. Steve Jordan, a biolo-

gist for the Maryland Department of Natural Resources and director of the Oxford Cooperative Laboratory, has suggested that one of the strongest indicators of a healthy Bay system would be diversity of such bottom-feeding fish. Some species of bottom-dwelling fish, particularly sheepshead and sturgeon, almost have disappeared from the Bay. Oysters and clams have also declined precipitously in recent years. Since the early 1600s, the diversity of diatoms—nutrient-rich algae that are significant to the diet of many aquatic species in the Bay—also has decreased. The reason for this failing index of the Bay's health is that, since 1608, the Bay's bottom has been laced with sediment, toxic substances, dead algae, and other contaminants dumped there.

A Bay restored to the conditions of the early 1600s would have more bottom-dwelling fish, oysters, and other benthic species than it does now.[10] These species feed on excess algae in the Bay, thus improving the Bay's water quality. Oysters, which traditionally have completely filtered the Bay's water, will not soon return in sufficient numbers to accomplish the job. In the meantime, the Bay needs filtering by other means, from the black filter strip fencing that surrounds construction sites to the continued restoration of aquatic grasses and other shellfish equally capable. Restoration of the Bay's health may not mean restoration of conditions exactly as they were in 1607. As an EPA official explained: "Because we cannot return to the way nature gave us the Grand Canyon and the Bay is not a reason to fight any less for their preservation as healthy and diverse ecosystems. A restored Chesapeake will be a different bay than it has ever been, but it can still be as healthy, as diverse and as compelling a place."[11]

This nonnumerical description of the characteristics of the Bay has stressed openness, variability, and complexity. These concepts serve as prelude to our historical study. The Bay is also open in a political and economic sense. People's use of the Bay is very much affected by events taking place outside the region. The recent buildup of demand for coal-shipping facilities, a result of the world energy market, is a case in point. The ports of Baltimore and Norfolk export coal to the world as they have for 150 years. Through this funnel of a bay pour products from the hinterlands of Virginia, Maryland, West Virginia, Pennsylvania, and parts of New York and North Carolina. Today these two ports are among the top five in the United States for volume of trade, surpassed on the East Coast only by New York. This situation must be kept in mind as we study Chesapeake history.

We learn the importance of facts when we realize how the variability of the Bay's quality produces popular misunderstandings. A large fish kill (or, more appropriately, a fish die-off) is typically taken as prima facie evidence of pollution. To argue without other evidence that there is a direct causal link, however, is to commit the fallacy of *post hoc, ergo propter hoc* (after this, therefore because of this). This line of argument, regardless of what some ancient Greek logicians might think of it, has been often employed on the Chesapeake.

3

Many Deep-Water Ports

---

*It is called the Bay of the Mother God, and in it there are many
deep-water ports, each better than the next.*

—Brother Carrera, 1572

This first European record of the Chesapeake also described the
Bay as great and beautiful.[1] The year was 1572, and the writer,
Brother Carrera, was a Spanish priest. But it was the gleam of commerce in Carrera's eye that produced fateful consequences. New-
comers to the New World arrogantly claimed it by "right of discovery." In their minds bloomed the image of wealth, of riches greater
than Spain's in Latin America. The chief exploiter of this dream, Sir
Walter Raleigh, hired Richard Hakluyt to write *Discourse on Western
Planning* in order to convince Queen Elizabeth and her counsellors to
provide a lot of royal backing to support settlements on a large scale.
This work pointed out advantages to the prosperity of the country:
"By makinge of shippes and by preparinge of thinges for the same, by
makinge of cables and cordage, by planting of vines and olive trie,
and by making of wyne and oyle, by husbandrie, and by thousandes
set on worke."[2]

This visionary book imagined towns "upon the mouthes of the
greate navigable rivers." And Raleigh then proposed his second settlement in the New World, to be established on the Chesapeake Bay.
On January 7, 1587, it was incorporated by the governor and the "As-
sistants of the City" of Raleigh, Virginia. Raleigh's propaganda and
overtures inspired enthusiasm in others, although his own plans
came to naught.

The next move towards the Bay was better planned. The Council
for Virginia of the London Company spelled out the moves for the
new colonists. Among the instructions, one provided that colonists

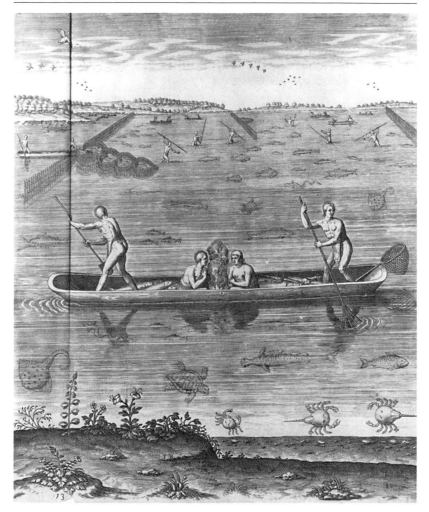

A late–sixteenth-century scene painted by John White. Thomas Harriot in 1590 noted that the Indians have many fish of kinds never found in Europe, and all of an excellent taste. Of the Indians, he said, "They are untroubled by the desire to pile up riches for their children . . . sharing all those things with which God has so bountifully provided them." Courtesy the Library of Congress.

must not "plant in a low or moist place because it will prove unhealthful." Unfortunately, the colonists did not heed this advice when they chose their site.

Much has been written about the unhappy years after the choice of the hundred-odd settlers to settle in Jamestown in the spring of 1607. One historian has conjectured that the colonists poisoned themselves with salt from the brackish waters and infected themselves with pathogens from their own wastes.[3] This detective work in public health may or may not be accurate, but it is certainly the case that by October of 1608 over one-half of the hundred-odd newcomers had died.

Not all experiences proved bad. In April 1607, one colonist reported of Lynnhaven Bay that "we got good store of mussels and oysters, which lay on the ground as thick as stones. We opened some and found in many of them pearls." That same month at Cape Henry colonists broke up an Indian oyster roast and pulled from the fires oysters "very large and delicate in taste." Two months later a letter from "Council in Virginia to Council in England" told of a Virginia river notable for "sweetness of water" and added that it was "so stored with sturgeon and other sweet fish as no man's fortune has ever possessed the like."[4]

After the first winter at Jamestown, Captain John Smith led a few men in a shallop on a voyage around the Chesapeake. The Bay proved to be a monster. Its winds, rains, mosquitoes, and Indians tested the man famous for bravery, boastfulness, and energy. He mapped the Eastern Shore as best he could. But the complex interweaving of water and land defied him, observant though he was. Smith Island recalls his visit in its name.

On the western shore he ascended the Potomac a ways and recorded on his map what Indians reported of its course beyond his range. He discovered the Patuxent River and the cliffs of Calvert, which he called "Richard's Clifts." On the way north he missed the West, Rhode, South, Severn, and Magothy rivers, but found the Patapsco. Evidently he penetrated what is the Inner Harbor of modern Baltimore and noted the hill later called Federal Hill: "For the red clay resembling bole Armoniac, we called it [the river] Bolus."[5] At the head of the Chesapeake Bay he was disappointed that he failed there to reach the Indies. The year before he had read instructions from the London Company about choosing a navigable waterway running well into the country, preferably in a northwesterly direction, "for that way you shall soonest find the other sea [that is, the Pacific Ocean]."[6]

TOBACCO TAKES HOLD

Three years later, in 1611, the London Company and its settlers received a gift richer than the gold of the Indies that Smith had sought. It came with the immigrant John Rolfe, imported from the West Indies, and it was, of all things, a simple plant—tobacco. This plant flourished in the leaf mold–rich soil of Virginia. Within a decade everyone there was planting and harvesting it.

Back in 1586 Sir Walter Raleigh had received a tobacco pipe by way of Spain. Shortly afterwards doctors latched onto tobacco as *herba panacea*, a cure-all. Other people called it *herba santa* or "devine tobacco." The upper classes of the Old World took it up. And Virginia could—and did—feed the habit. With the news of how much money could be made growing tobacco there, the colony attracted hordes of English people, most of them unfit to live on a frontier. Still they came, and by 1640 all the best land on the James River up to the falls (the site of Richmond) had been taken up.

Evidence of the appetite for arable land appears in an advertisement in Williamsburg's *Virginia Gazette* on August 29, 1771: "A plantation for sale in King and Queen County, just below West Point, known by the name of Gough's Point, containing two hundred acres ... on which is a good Deal of Marsh that with little Expense might be made a valuable Meadow."

Virginians, however, soon had competition. Although their company had received three successive charters from the English Crown, their boundaries in Virginia proved amorphous to the king. The terms of the three documents pushed the boundaries out to include what are now parts of North Carolina, all of Maryland, and parts of Pennsylvania. Then on June 20, 1632, King Charles I granted to Lord Baltimore the tract of land that Baltimore called Maryland. Who could question his majesty's absolute right to give away the same Bay twice?

Here then arose a second colony, and one whose settlers, first arriving on its shore in 1634, sought gold too. They came to an inlet off the Potomac, St. Mary's River, and called the settlement St. Mary's City. That name seems ironical to us because of its expectation of urban distinction in the tidewater where no real city spread until 175 years later. St. Mary's City turned out to have a better site than Jamestown because it stood above its river and was well supplied with fresh water.

Colonial rules, set by proprietors and kings, encouraged the founding of port towns. And the authorities chartered many all around the Bay. They and their merchants clearly hoped for new versions of London and Bristol. In 1705 an advertisement appeared in London called "A Plain and Friendly Perswasive to the inhabitants of Virginia and Maryland for promoting towns and cohabitation": "Cohabitation would not only employ thousands of people . . . others would be employed in hunting, fishing, and fowling, and the more diligently if assured of a public market."[7]

The planners concentrated on towns as conduits for the money that was to flow into public and private coffers. What they got instead were headaches. The first Bay settlement, Jamestown, according to ample evidence, seemed always about to collapse. But government officials persisted in authorizing the establishment of towns, even though most efforts failed. When an earlier Baltimore, laid out on the banks of the Bush River in Harford County, Maryland, did not develop, it was replaced by another Baltimore in 1729 on the present site.

Again, the Maryland colonial assembly took great pains in 1744 to establish Charlestown (in Cecil County on the Northeast) by giving special privileges to owners of waterfront lots. The assembly offered use of a town wharf and warehouse, free use of public squares, spring and fall fairs, and permission to lease marsh areas in town to anyone who would drain them. Similar lures succeeded better at Georgetown, Maryland. The commissioners for the new District of Columbia published regulations on July 20, 1795, permitting proprietors of water lots to build wharves "as far out as they think proper [but not injuring the channel]."[8]

Although public policy encouraged urban development, no town had developed by 1662 in the Virginia colony, a place that then had about forty thousand population.[9] The reason for this failure lay in the economy of tobacco growing on far-flung fields of plantation owners who shipped from riverside docks nearby. No planter required a town. But English colonial officials, never facing up to the geographic and agricultural realities of the Chesapeake region, wanted Virginia as well as Maryland to be like New England, a town-based culture. A contemporary comment illuminates the problem: "No country in the world can be more curiously watered. But this conveniency that in future Times may make her like the Netherlands, the richest Place in all America, at the present I look on . . . as . . . the greatest Obstacle to Trade and Commerce."[10]

Chesapeake residents did not need cities then. What they did need was land, lots of tobacco-growing land. And they needed waterways to carry the pungent-smelling cargo abroad. They found both land and water aplenty. Father White, accompanying the first Marylanders, called the Potomac River "the greatest and the sweetest I have ever seen. The Thames is but a little finger to it."[11] There in southern Maryland the settlers made a good beginning. They avoided the strange fluxes and agues that had afflicted the Virginians earlier. And they enjoyed the abundant game and fish. They learned from the Indians how to grow maize and sweet potatoes. Because the act of religious toleration admitted Quakers, some of that industrious sect had settled around the Bay by 1650, particularly in what is now Anne Arundel County. In 1649 a group of Puritan exiles from Virginia founded Annapolis (at first called Providence), and that settlement became a port of entry. Thirty-five years later it was made the seat of royal government, the proprietorship of the Calverts having ended.

Annapolis did not become much of a town, although as a colonial capital in the eighteenth century it had its pretensions like Williamsburg, the second capital of Virginia (1699). We have but to look at population figures. Writing in 1782, Thomas Jefferson reported that Williamsburg never exceeded a population of eighteen hundred. Annapolis had about the same. Richmond, Virginia's third capital and unincorporated until 1805, had only eight thousand in 1820, but by that date Baltimore had zoomed ahead of all tidewater towns (for reasons that we will explore) to sixty-two thousand.

Baltimore stood out as a port as early as 1785, when a visitor noted, "Alexandria [Virginia] has made considerable advances since 1778, but afforded no comparison in its progress, to its vigorous rival, Baltimore."[12] Even moving the new nation's capital to the shores of the Potomac in 1800 failed to create much of a city there for years. Note that the largest marsh in the District of Columbia was reclaimed only after 1900. It occupied part of the Mall.

The reason for the small size of the towns rested on tobacco. Back in 1640 half of the English settlers in America lived in the Massachusetts Bay Colony. Of the other 15,000, about 8,000 lived in Virginia, and 1,500 in Maryland. Twenty years later, 26,000 whites and 950 blacks lived in Virginia, and 7,600 whites and 750 slaves lived in Maryland. During the next fifty years the number of slaves increased astronomically, because as many as possible were needed to grow tobacco.

In 1724, an Englishman, the Reverend Hugh Jones, published in his *Present State of Virginia* an observation that applied to Maryland as well:

> No country is better watered, for the conveniency of which most houses are built near some landing-place; so that anything may be delivered to a gentleman there from London, Bristol, etc. with less trouble and cost, than to one living five miles in the country in England. . . . Thus neither the interest nor inclinations of the Virginians induce them to cohabit in towns; so that they are not forward in contributing their assistance towards the making of particular places, every plantation affording the owner the provision of a little market.[13]

The "little market" resembled a company town, with the head of the company, the planter, then fulfilling many roles: farmer, judge, merchant, doctor, host, and—always—aristocratic lord of the manor in the English mold. We do well to look at him closely. He was at the center of economic, social, and political events in the tidewater. In him we find influences on the governing of uses of the Bay. He served as legislator, he acted as sheriff, he decided as judge.[14]

Economically speaking, the planter and his tobacco crops dominated the Bay region. These Virginia and Maryland planters ruled over kingdoms. George Washington owned a mere 8,000 acres, small compared with the 63,093 acres of Robert Carter, of Nomini Hall, Virginia.[15]

With 5,000 acres and the work of slaves, a planter could afford to live like a lord. (Even 50 acres of tobacco land supported a family in the eighteenth century.)[16] With money to live like a lord, the planter did ape the social life of the English gentleman of status and large estate. And like that gentleman, he conveyed his estate to his heirs to ensure keeping the land in the hands of a few. On his land he built and maintained a great house. With plenty of credit in London, he bought luxuries to adorn it. He also bought culture. Though the level

Tobacco Road, Richmond, 1845. A WPA-sponsored mural in a series about Richmond, Virginia. Such roads, called "rolling roads," were found in many communities around the Bay—we still have many Rolling Roads. Courtesy the National Archives.

must have been superficial at first, the culture of the tidewater deepened by 1750 to shine in what has been called the golden age of the Bay aristocracy.[17]

THE COMPACT OF 1785

That golden age shaped the major founding fathers who joined in the creating of the United States, including Washington, Jefferson, Madison, and Monroe. Any Bay history with a humanist approach, then, must record the leadership of planters like these men. George Washington exemplifies the early Virginians' way of settling regional conflicts. Note that he too was an English gentleman, willing to accept change but possessed of traditions that became the foundation of the United States. He was loyal to his parish and to his Chesapeake fiefdom. No wonder he presided over a constitutional convention that codified the federal spirit and also expressed support of states' rights.

It is significant for our present history to note how Washington sat down with a handful of other Chesapeake men in Alexandria and at Mount Vernon to solve major conflicts about Chesapeake waters that had long disturbed Marylanders and Virginians. He was not yet president when he and the group worked out the historic Compact of

Washington boasted that "no estate in United America [was] more pleasantly situated" than his Mount Vernon. In this painting by Benjamin Henry
Latrobe (the only known picture of Washington drawn from life in this setting), Washington is shown scanning Potomac shipping with a spyglass.
Seen here in July 1796, the Washingtons are entertaining Lafayette's son,
seated next to Mrs. Washington and the tea urn. Also present are Nellie Cus-
tis, a granddaughter, who stands posed like Flaxman's Helen of Troy from a
contemporary edition of the *Iliad*, and Master Lear, the son of Tobias Lear,
Washington's secretary in charge of affairs at Mount Vernon. Latrobe noted
that "down the steep slope, trees and shrubs [are] thickly planted but kept so
low as not to interrupt the view but merely to furnish an agreeable border to
the extensive prospect beyond." Courtesy James W. Tucker.

1785. Their decisions freed Marylanders from the threat of being
forced to pay tolls to Virginia at the entrance to the Bay. They also
gave Virginians the long-sought right to fish the Potomac River,
owned entirely by Maryland.

In another historic decision but one of less lasting duration—one
hundred years—this compact moved the boundary between two
states. The line set in Lord Baltimore's charter of 1632 had begun well
south of the mouth of the Potomac and below South Point on Smith
Island. The new line gave much more of the Bay to Virginia and included what is reported as 20,000 acres of oyster bars.[18]

Chesapeake traditions behind this compact reached through it to influence the whole nation. Because the meeting of regional leaders at Alexandria and Mount Vernon had gone so well, the committee members decided that other quarrels between states could be settled by bringing together delegates from all thirteen states. The first call sent out by Virginia failed to bring a big enough delegation to Annapolis. But the next call was to meet at Philadelphia in the summer of 1787. Out of that meeting came the U.S. Constitution and the selection of Washington as first president of the new United States. More than a few historians trace modern American democracy back to the Chesapeake Bay and not to New England.[19]

THE RISE OF BALTIMORE

Even when commerce went to town, leadership stayed in the country. But in the nineteenth century towns did grow, and the story of the growth of Baltimore, the largest of them, will illustrate how and why.

Baltimore first made its reputation as a brash town on the Chesapeake Bay. Other tobacco ports, of course, existed before the chartering of Baltimore in 1729. This new town shipped little tobacco as it spread across land owned by the great Carroll family and in time surrounded the plantation house of Charles Carroll the Barrister, Mount Clare. That mansion, which still stands, received guests like Washington and Lafayette in the eighteenth century.

Those early guests stopped at Baltimore as others less lordly did because Baltimore grew up at a crossroads: It happened to have a harbor on the main overland route to points south from Philadelphia. More significantly, like Richmond, Virginia, it stands on the fall line of its river. The significance of its position lay in the rush of water in streams, or falls, as they were called, that powered mills. Probably the first mill was erected by Jonathan Hanson about 1710, although in 1661 David Jones bought 380 acres of the stream named for him, Jones Falls, and may have built mills there. A century later, in 1805, the area within a 20-mile radius of town held sixty mills. From them poured so much flour that that region held the world record. Clearly the economy of this piedmont region then rested on wheat and not tobacco.[20]

Additional reasons for the rise of Baltimore lay in the use of nearby waters for shipping and shipbuilding. An English Quaker

View of Baltimore from Federal Hill (1850) by E. Whitfield. At this time Baltimore ranked as one of the largest cities in the nation—and one of the busiest, as the number of smokestacks and ships shows. At the far right a forest of masts marks Fells Point, shipbuilding center and the original deep-water port of Baltimore. Courtesy the Library of Congress.

family named Fell, one of them a shipbuilder, bought land on deep water (now Fells Point) and inaugurated an industry and a port. Named for an Irish port, Baltimore had received immigrants happily, and one of them from the north of Ireland, Dr. John Stevenson, often receives credit for the idea of shipping the good Maryland flour abroad. Trade burgeoned. By 1800 Baltimore had become the gateway to the West, the South, the Caribbean, and Europe.[21]

Still, Baltimore was a city-come-lately among East Coast ports. It had only five thousand population in 1770, when Philadelphia ranked as the second-largest town in the British Empire. No wonder Congress complained about Baltimore when, in 1776, it took refuge on Baltimore Street after the capture of Philadelphia by British troops. The little town's shipyards, nevertheless, produced sturdy warships for Congress.

After the Revolution the fast schooners of Baltimore—the Baltimore clippers—carried its name around the world and brought wealth and crowds of immigrants. The town's privateers created the "Nest of Pirates" that the British feared and sought to destroy in the battle of Baltimore at Fort McHenry, a climax of the War of 1812. The citizenry rose as one to rebuff the attack in September 1814. They also sang about the victory in a song, "The Star-Spangled Banner," written by one of them and they named streets after its heroes. To this day Baltimore remains the only major American port never occupied by a foreign power.

With the creation in 1828 of the Baltimore and Ohio Railroad in Baltimore, grain came from the Midwest to be distributed around the world. But grain and flour are only part of the lucky combination that made Baltimore second only to New York in size and importance by the time of the Civil War. Another reason for growth was the metal industry. Back in the eighteenth century, enterprising men like the Carrolls, using local deposits, developed one of the largest iron-production facilities in the British Empire. Entrepreneurs also built industries from local copper and chrome deposits.

Clearly the citizens of the town had guts as well as enterprise. They created first the clipper ship and then the Baltimore and Ohio Railroad. Both modes of transportation helped double the population every decade. Chief molders of the city's character appear to have been businessmen, of the Venetian stamp,[22] as one memorialist called them. Shrewd and aggressive, they ruled the world trade routes as well as the city. Yet part of the city's character yielded to the civilizing waves of gentility and genius—and of women—that washed over the town between 1800 and 1860. It makes a town special to have held within its bounds the first American saint, St. Elizabeth Ann Seton, as well as artists of the caliber of the architect Benjamin Henry Latrobe, the Peale family of artists and scientists, and Edgar Allan Poe.

A less attractive side of the city's character caused Baltimore to be called Mobtown. At Commerce and Pratt streets was spilled the first blood in the Civil War, when a mob attacked troops from Massachusetts who were rushing to defend Washington. Baltimore remained divided between North and South, a price paid for its location. Until the end of the war it felt what the state's anthem called the despot's heel against its windpipe.

After the Civil War, trade in the port of Baltimore still benefited from the harbor being 150 miles nearer the West Indies and South America than New York. Among the imports were bananas, sugar, spices, and coffee. Baltimore had its own "coffee fleet." Guano came in to be used in fertilizer, a product that also used local oyster shells in its manufacture. These shells came from oyster canneries that lined the northern waterfront. The canning of oysters developed in Baltimore partly because of abundant supplies nearby, and partly because of cheap labor in the form of both European immigrants and freed blacks. Another reason lay in the ingenuity of a number of inventors, including Thomas Kensett, who invented the machines for making and sealing tin cans.

Before 1900 Baltimore had became the nation's largest canning center. Processing other products of the Bay region—vegetables and fruits shipped to Baltimore docks—pushed the industry so much that by 1890 canning employees numbered twenty thousand. Remaining from that era are companies like Crown Cork and Seal and Continental Can that grew up as auxiliaries to the canning factories.[23]

Let us pause here at the economic successes of the Bay's canning industry that sent its products around the world, and look back over three hundred years. In 1607 the Chesapeake held out false promise as a route to the Far East. Settlers followed the disappointed explorers to create a tobacco aristocracy dispersed along its shores. Later came the port cities, preeminently Baltimore, to serve as points of transshipment. All the while, settlers and their successors exploited the waters of the Bay region. For them the quality of water was not a problem, but a solution. Besides shipping, water was used for drinking, as a sink to dispose of wastes, and as a fishing ground. The problems arose when these uses came into conflict.

4

A Well, a Sink, a Pesthole

---

*The Back Basin [of Baltimore in the 1890s] received the effluence of
such sewers as existed and emitted a stench as cadaverous and
unearthly as that of the canals of Venice. . . . It continued to afflict the
town until the new sewerage system was completed, and the Back
Basin was reduced to the humble status of a receptacle for rainwater.*
— H. L. Mencken, 1941

The settlers on the shores of the Chesapeake first demanded pure
water to drink. Not all the imported beer and sack could substitute for it. Planters and townsmen alike had trouble finding and retain-
ing sources of good water. As with so much else, we can begin with
Captain John Smith. On the Eastern Shore at one point in his explora-
tions of the Bay, he and his crew found only a puddle from which to fill
"but three barricoes. . . . We digged and searched in many places, but
before two daies were expired, we would have refused two barricoes
of gold for one of that puddle of water of Wighcocomoco."[1]

OBTAINING AND PROTECTING A WATER SUPPLY

When springs were lacking, the inhabitants dug wells. They also dug
latrines and, later, cesspools. Luckily, the hilliness of Richmond and
Baltimore permitted good draining of storm water and of domestic
and industrial wastes up to a point. Also, for years Baltimore's lusty
streams rushing through the hills provided both drinking water and
a channel for wastes to enter tidewater. This mixture of uses inevitably led to trouble.

It led, moreover, to attempts not only to provide a supply but
also to protect the drinking water in towns. In 1792, for example, the
Insurance Fire Co. of Baltimore was authorized to sell shares to obtain capital for supplying the town with a reservoir and with pipes to
conduct the water.[2] Again in 1800 the state legislature gave Baltimore

permission to construct a water supply system. We must keep in mind that Baltimore's population was just about doubling every decade after the American Revolution. By 1809 the town's natural springs were being encroached upon by this growth, and "threatened with destruction by other improvements."[3]

Central Springs on North Calvert Street (the present site of Mercy Hospital) was incorporated, as soon were two others, Clopper's in the south of town and Sterret's in the east. Like the water company in New York City (from which evolved Chase Manhattan Bank), the Baltimore Water Company (established in 1808) became more intent on other money-making ventures than on providing streams of clean water. The January 15, 1830, *Baltimore Gazette and Daily Advertiser* carried a report of the town's water committee asserting that that company provided only for the central section of town. Besides, ran the complaint, more than half the year the color and consistency of the water made it unfit for any use. And sometimes there was no water at all. This news report went on to say that the water company was "chaining down our fellow citizens to a prolonged subsistence on this muddy substitute for the pure element." People lived in fear of resorting to "pumps which we have found it necessary to abandon, and imbibe disease and putrid water together." Almost as bad as disease was the threat of having to "relinquish the design of multiplying our houses, our streets, and our people." For a booming town that threat must have hit hard.

Norfolk fared even less well. From its earliest days visitors noted that the water was brackish and, some said, unpalatable. A public spring was located near Main and Church streets, but the river itself provided "the place appointed [by the Norfolk Council] for the public laving." So bad was the taste of the drinking water from pollution and brackishness that in the summer of 1800 a man who owned a good well on Briggs' Point, Johnny Rourke, peddled water from his "tea-wagon." Fifty years later most of the drinking water came from cisterns. And only after the Civil War did the city begin to tap Lake Drummond and Deep Creek for good, pure water.[4]

Authorities took steps to prevent contamination of water supplies. In fact, an early use of the word *pollution* appeared in a provision of an act of the Maryland legislature establishing the Baltimore Water Company in 1808.[5] A fine was imposed on anyone polluting the Jones Falls, the source of drinking water, between the pumping house and the mill. This stream evoked plenty of controversy as a source, partly because its supply ranged from trickle to flood. The Baltimore Water Commission in 1852 rejected a report urging the use of the Gunpowder River as a source. Instead, still drawing on the Jones Falls, it built a system complete with classical temples disguising wasteweirs and gatehouses. But their work was in vain, for the supply proved inadequate. Worse still, the absence of filtration meant contamination and typhoid epidemics. Soon came new waterworks drawing on the Gunpowder River. Ultimately, Baltimoreans were compelled to send 50 miles to the Susquehanna River for water.

Various government officials in Maryland recognized throughout the latter half of the nineteenth century that public water supplies were being contaminated by materials thrown into waterways. The report of the Baltimore Water Commission in 1853 argued against building an open canal to the city's reservoir from the source, because "water is exposed to all the filth and dirt which anyone chooses to throw into it."[6] In 1874 it was made illegal to throw dead animals and other substances into the Potomac River.[7] That tributary provided drinking water for many settlements, including the District of Co-

During the nineteenth century, public springs and corner pumps supplied water in Chesapeake towns. This pump stood at the corner of Gorsuch Avenue and Loch Raven Road, Baltimore, as late as 1936. Courtesy Enoch Pratt Free Library.

lumbia. As late as 1886 the Maryland legislature had to reiterate that there was pollution of the water supplies.[8]

A classic case of conflict about the use of water for drinking and for disposal of wastes was decided in 1882 when the mayor and city council of Baltimore sued the Warren Manufacturing Company.[9] The defendant's cotton factory, it was complained, dumped "divers injurious ingredients and substances" into the Gunpowder River, which had been dammed for reservoirs by Baltimore city. The resulting pollution contaminated the drinking water of Baltimore. The court held that the privies and hogpens near the factory did indeed pollute and had to be removed. But the city was denied relief against the manufacturer since it was unable to prove the nature of the pollutants emanating from the manufacturing processes of the company.

A sampling of the case law in Virginia during the nineteenth century reveals that the courts had to deal mostly with mills and milldams. Courts decided not only issues involving competing rights, but they also addressed pollution issues. For example, in 1828 a milldam across Little Creek in Nottoway County, Virginia, caused stagnant waters. The commonwealth sued, alleging that the dam was a public nuisance. The plaintiff lost, having failed to show how stagnation affected a public highway or some other place in which the public had such special interest.[10]

In 1833, a court dealt with another case of stagnation caused by the construction of a mill and milldam, this one on Deep Creek in Virginia. The Maud family suffered "severe bilious diseases" every year, and several "intelligent physicians" and others from the territory testified that the diseases were "justly imputed to the effluvia from the stagnant waters of the mill pond." After the washing away of the mill, a suit was brought to prevent the owner from rebuilding. The latter argued that he had built under statutory authority and had a statutory right to rebuild. The court of equity asserted it had the authority to "prevent the destruction of health and life," and it distinguished between danger to health and danger to fish and navigation. The court ruled that it would allow experiments with the latter, but, as to the former, "The law tolerates no such experiments at the risk of human life."[11]

Finally, at the close of the nineteenth century, the Virginia Supreme Court of Appeals dealt with the problem of pollution itself. In *Trevett v. Prison Association of Virginia*[12] the court ordered that damages be paid to the plaintiff, who had used the waters of a stream for domestic purposes and for a number of cows whose milk and butter commanded the highest price on the market. The defendant operated a school for young criminals which emptied "refuse water, urine, and excrement" into the stream, thereby destroying the plaintiff's dairy business.[13]

CONTAGIOUS DISEASES

Health certainly posed the major issue connected with water quality in the first 275 years of recorded Bay history. The chief enemies to good health—contagious diseases—were reported in newspapers every August. Epidemics as well as individual incidents filled letters and journals with appropriate alarm. For example, in the fall of 1762 Charles Carroll the Barrister wrote from Annapolis, "I this summer made an Excursion as far as Boston in order to Escape my Troublesome annual visitant the fever and Ague [possibly a form of malaria] but had not been returned to Annapolis four days before I was seized with it in a more violent manner than at any of its former attacks and it still keeps possession of me."[14] In another instance, Colonel Landon Carter recorded in his diary for October 3, 1765, "This is a strange ague and fever season. The whole neighborhood are almost every day sending to me."[15] These neighbors no doubt were sending for medicines and advice.

In 1797, however, Dr. John Davidge, a leader in Maryland medical circles, had declared that yellow fever was not contagious. In Philadelphia the well-known Dr. Benjamin Rush agreed, but few others did. People noticed that this disease always started at Fells Point, usually near Smith's Dock. Some noted that just east of the point and between it and the lazaretto stood stagnant ponds. That was true during the epidemic of 1819.[16]

After 1822, statistics show a decline in deaths in Baltimore as yellow fever and malaria diminished. The reasons lay in better drainage and in less commerce with the West Indies. Confirmation comes from an authority on water quality, Abel Wolman of The Johns Hopkins University, who wrote in 1969:

The effects of epidemics in the early days of the Republic on pushing forward installation of public water supply and sewerage facilities can hardly be overestimated. Yellow fever and cholera between them created havoc decade after decade with important results so far as municipal housekeeping was concerned. . . .

The increasing public demand for sewering and draining cities to avoid nuisance and to gain convenience diminished remarkably the incidence of yellow fever, due primarily to the elimination of the breeding places for mosquitoes. In similar fashion the epidemics of cholera in 1832, 1843, 1849, 1854, 1865 and 1873 brought about a certain amount of recognition of the probable relationship between this disease and the inadequate and unsafe methods of the disposal of household wastes, and undoubtedly stimulated the establishment of drainage and sewerage systems in the larger communities.[17]

However, for centuries people had believed that contagious diseases were airborne. The truth did not come until the very end of the nineteenth century. Appropriately for our history, two Chesapeake doctors helped fix blame on the water. Although the bacterial cause of cholera wasn't proven until 1884 in Egypt, a generation earlier a Baltimore doctor came close to the truth.[18] Dr. Thomas Buckler attempted a scientific study of a cholera outbreak in the almshouse at Calverton outside Baltimore as well as within the city. Then in 1851 he published *A History of Epidemic Cholera*. In it he cited a case of filth in Baltimore in connection with nine deaths from cholera.

The other Baltimore physician, Dr. Jesse Lazear, in 1900 became a martyr to proving the link between mosquitoes and both malaria (swamp fever) and yellow fever. A fellow worker on the United States Army Yellow Fever Commission, Dr. James Carroll, wrote that Lazear contributed "the two first authentic cases of experimental yellow fever on record. . . . Then with a full knowledge of the power of the insect to convey the disease, he afterwards calmly permitted a stray mosquito that had alighted upon his hand in a yellow fever ward, to take its fill, and inject into his system the virus that twelve days later robbed him of his life."[19]

Today an inscription on a plaque in Johns Hopkins Hospital (where Lazear was educated) states: "With more than the courage and devotion of the soldier he risked and lost his life to show

Four scientists, all born in the Chesapeake region in the nineteenth century, who have made lasting contributions in the field of public health. *Above,* Dr. Thomas Buckler in 1870, an early investigator of the cause of cholera. (His image is superimposed on what at that time was considered a suitable background for a doctor.) Courtesy Maryland Historical Society. *Below,* Dr. Jesse Lazear, a native of Baltimore who trained at Johns Hopkins Medical School. Lazear served on Dr. Walter Reed's Yellow Fever Commission, let an infected mosquito bite him, caught the fever, and died. His death, on September 25,

1900, helped prove the cause of the disease. Photo taken in the 1890s. Courtesy The Alan Mason Chesney Medical Archives, The Johns Hopkins Medical Institutions. *Above,* Dr. William H. Howell in 1900, a founder of the Johns Hopkins School of Public Health, the first school of its kind in the world. Courtesy The Alan Mason Chesney Medical Archives, The Johns Hopkins Medical Institutions. *Below,* Professor Abel Wolman of Johns Hopkins University, author and internationally recognized authority on water quality, who was still working at age ninety. Photograph by William C. Hamilton. Courtesy Ferdinand Hamburger, Jr., Archives, Johns Hopkins University.

how a fearful pestilence is communicated and how its ravages may be prevented."

The summer epidemic of 1855 in Norfolk well illustrates the horrors of yellow fever. A steamer bound from St. Thomas to New York put into Hampton Roads for repairs. On board were cases of yellow fever, a fact kept hidden until a shipyard laborer contracted the disease. By August the city of Norfolk as well as Portsmouth suffered decimation of population. Reports said that the city was wrapped in gloom. Most people were either confined with sickness or waited on the sick. The number of deaths rose to one hundred daily, and the supply of coffins gave out. The calamity shocked residents, particularly since they had begun to believe that their attention to draining streets and using cisterns for drinking water had saved the city from the recurrence of epidemics such as those of 1821 and 1836.[20]

In 1878, Congress passed the National Quarantine Act. This and other such acts gave to federal agencies some duties that for almost three hundred years had been performed by local governments. For example, as early as 1801 the Maryland General Assembly had authorized the construction in Baltimore of a lazaretto at the northeast edge of Baltimore harbor.[21] This city already had its health police, working for the city's appointed commissioners, that was set up when Baltimore was incorporated by the state legislature in 1797.

About the time of the National Quarantine Act, states were establishing official boards of health. Maryland's was created in 1874 to guard the sanitary interests of the people of the state, to investigate the causes of disease and mortality and the influence of locality, employment, habits, etc., on health,[22] and to investigate nuisances. Local boards of health were authorized by Maryland's legislators in 1886.

Just four years earlier, in 1882, Dr. W. C. Van Bibber read a paper before the Medical and Chirurgical Faculty of Maryland, later published as "The Drinking Waters in Maryland Considered With Reference to the Health of the Inhabitants."[23] He reported that on January 19, 1881, the hydrant water of Baltimore had tasted and smelled "disagreeable," and that the public was aroused. "The public-spirited proprietors of the *Sun* newspapers" hired a Professor Tonry, and the water board got Professor Ira Remsen to investigate the cause. Tonry gave his opinion that the source, Jones Falls, was polluted "due to the decomposition of the sulphates, held in so-

The original lazaretto (now demolished), built by Baltimore in 1801 as a contagious diseases hospital and quarantine station. In the foreground are navigational buoys brought in for refurbishing in 1940 when the building was used by the Coast Guard. Photograph originally appeared in *The Baltimore Sun*. Courtesy *The Baltimore Sun*.

lution, passing into the sulphites, and setting free sulphuretted hydrogen gas." The bad taste and smell lasted until the middle of April. Citizens using pump water instead of hydrant water were warned by Dr. Van Bibber to be suspicious of their supply also. "When it is known that there are over 80,000 nuisance sinks and wells at various levels within the city limits of 9,600 acres, no one will wonder that the pump waters within this area, and even beyond it, should be contaminated."[24]

Just how far attitudes had changed we can see by comparing these steps with the comment of a civil engineer, Charles Varle, writing only fifty years earlier in 1833. After praising what he called the efficiency of the health police of Baltimore, he noted the calamitous ravages of cholera in 1832: A total of 3,572 people had died out of a population of 80,990. Varle then added, "May the great Disposer of events hereafter keep us free from pestilential visitations."[25]

ELIMINATING MARSHES AND FOUL ODORS

Less pious citizens prodded early officials to prevent the spreading of disease associated with water and waterfronts. The fever, it appears, was more terrifying than fire or flood. One gadfly, Dr. Thomas Buckler, published a map in 1851 showing the areas of remittent and intermittent fevers around the Patapsco River, which were extensive. Buckler had his theories of the causes, just as other people did: He blamed uncleanliness generally and decaying refuse and garbage in particular. Darkly he prophesied that "the germs of an epidemic introduced into the polluted atmosphere surrounding this festering pond [the Inner Harbor of Baltimore]" would spread. His solution combined ingenuity with a money-making angle: simply shave off the hill adjacent to the basin, Federal Hill, and fill in the basin for building lots.[26] (Had the city fathers agreed with him, Baltimoreans today would have no Inner Harbor, the centerpiece of the city's renaissance.)

Thus Buckler would have rid the city forever of what he called "the most unwholesome, disease-engendering and pestilential condition of the Basin and Docks." He also would have eliminated "the stagnant mill-ponds along the Jones Falls, to say nothing of the putrid, seething and malarial condition of the tidal portion of the Falls." In addition, he suggested deepening the falls, so that "the tide may ebb and flow as high up as Madison Street bridge" and in its scouring remove the causes of disease.[27]

An urban geographer recently wrote that in the period of cholera epidemics from 1832 to 1873, a learning process was at work that resulted in what she calls the creation of "new institutions of collective response." The series of epidemics, she hypothesizes, called attention to certain "foul and filthy spots and spurred the increase of regulatory power."[28]

People particularly believed that the shorelines of harbors threatened disease. Public opinion, therefore, forced attention of town and county officials to get rid of marshes and low ground around settlements. In 1766 an "Act to Remove a Nuisance in Baltimore town" denounced in vivid language "a large miry marsh giving off noxious vapors and putrid effluvia."[29] Landowners were given four years to wall off this blemish on the waterfront and fill it in.

Early–nineteenth-century medicine sought the cause of diseases in the fumes given off from marshes and lowlands as well as from pu-

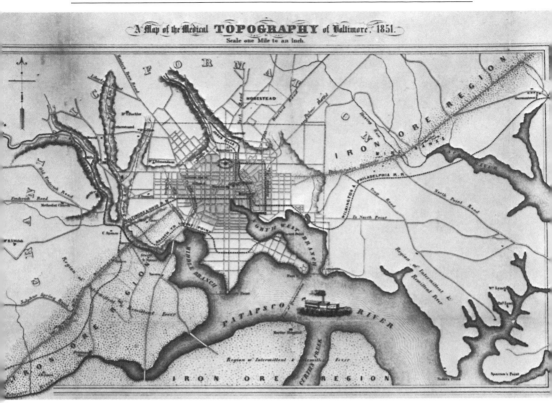

In 1851 Dr. Thomas Buckler published this map to show locations of the frequent outbreaks of fevers (by use of dark dots and shading). Note the reservoir, next to Jones Falls and north of the Washington Monument, which was the source of drinking water. The lazaretto can be seen on the map just above the steamboat. Dr. Buckler's house, marked in the upper left, was just north of the almshouse where he did his research. Courtesy the Library of Congress.

trescent vegetation. A contemporary historian concludes that Dr. Benjamin Rush "tried to improve the public health in Philadelphia by such common-sense expedients as sewage disposal, pure water, and clean streets."[30] In other cities similar analyses appeared. Among reports of the Baltimore Public Health Department for 1825 appears a note by a prominent doctor, Thomas E. Bond, who was called "Consulting Physician": the situation on Washington Street, Fells Point, "threatens to produce unfriendly effects on the health of neighboring inhabitants [because] the fall is insufficient to carry off the water with sufficient celerity. . . ." He added that below the apron [of the planks at the south end of the street] a "very dangerous nuisance" arose because "water . . . becomes stagnant, and emits at this season a very disagreeable smell and will in the hotter season of the year be a fertile source of deleterious exhaltation."[31]

So sensitive indeed were the noses of the nineteenth century that writers vied with each other to describe the various perfumes wafting up from the basin in Baltimore. One called it "hellbroth."[32] It was "a bubbling cesspool" remembered Abel Wolman, a man whose memory reaches back to 1900. He recalls returning at night in about 1910 on the boat from Tolchester and knowing when he was within five miles of the city by the odor.[33]

DISPOSING OF WASTE

The brackish waters of the Chesapeake Bay, while unsuitable for drinking, were an ideal sink for waste disposal. Inhabitants flushed their sewage and refuse into the receiving waters of the estuary. In Norfolk, wrote traveler Mederic Moreau de St. Mery in 1794, "The sewage ditches are open, and one crosses them on little narrow bridges made of short lengths of plank nailed to cross-pieces."[34] These ditches, of course, emptied into the convenient Bay.

Population growth compounded the waste disposal problems of cities. In the middle of the nineteenth century, Washington, D.C., appeared "as sour as a medieval plague spot," according to a later historian of the Civil War. She called the sanitary conditions of the crowded capital appalling. "With its river flats, its defective sewage system and its many privies, it had always been odorous in warm weather," but now in the wartime-crowded District of Columbia, epidemics of disease threatened. "The comatose Board of Health aroused itself to a protest, the Surgeon General made inquiries and the War Department conducted an investigation," she cynically concludes.[35] Not until 1889 were sewers constructed by the Board of Engineers.

In nineteenth-century Baltimore, the harbor ranked "among the great stenches of the world."[36] It stood out as the last of the great American cities to build a modern system for sewage, which came between 1906 and 1915. Writing in 1912, Clayton Coleman Hall boasted that completion of the system would make Baltimore "the best sewered and drained city in the world."[37] His enthusiasm carried him into error when he added that "the sewage [after processing] when discharged into Back River will be purer than the water into which it is discharged and will be absolutely innocuous to the oysters and fish inhabiting the waters of the bay." Such proved not to be the case.

Baltimore made a useful example of a city with initially defective sewage disposal after it became a metropolis of the Chesapeake region. By 1850 the city had grown to be the third largest in the United States, with 170,000 population. The seventy thousand represented its growth during the preceding decade. In an article about Baltimore in the contemporary *Gleason's Pictorial Drawing-Room Companion* (1855), the writer noted that "The city is built on quite uneven ground, which gives an advantage in relation to cleanliness."

Fifty years later, fifteen thousand houses had private lines to the Jones Falls for disposing of raw sewage. Streams like Harford Run and Jones Falls in Baltimore became nothing but canals to carry off the town's wastes. A member of Parliament, Sir George Campbell, commented in 1879, "A peculiarity of Baltimore is that there is no system of underground drainage" and in an understatement worthy of quotation added that this arrangement was "not very agreeable to the senses." But he was not certain that this city's old-fashioned way was not "much more wholesome than [London's] underground system."[38] Why, he didn't say. Possibly he was thinking of sewer gas. But more houses had cesspools that were cleaned regularly by Odorless Excavating Apparatus Company. The city's population, at that time half a million, was too large for what had once been of such great service, the good drainage of the hilly terrain and its streams. The soil became saturated, floods occurred in downtown, and the Inner Harbor smelled to high heaven. The critical Dr. Thomas Buckler proposed facetiously that the gases be collected to use in Congress. And he worried, not facetiously, that "other terms of an epidemic introduced into the polluted atmosphere surrounding this festering pond [will spread]."[39]

Dr. Buckler wasn't the only one offering proposals. In the mid-1870s fifteen plans lay on the table of the Joint Standing Committee on the Harbor. In a plan by Milo W. Locke, the water was to be kept in motion and diluted constantly. An editorial writer even seriously proposed building a dam across the mouth of the Inner Harbor. The water brought in at high tide could be held and released at low tide in

As late as 1915 pedestrians crossed open drains on raised stones in street beds. This engraving in the October 4, 1856, *London News* shows stepping-stones across Liberty Street, Baltimore. Courtesy the Library of Congress.

a flushing operation. The flood might, of course, wreak havoc on the ships below.

Another source of pollution came from the draining of filth from streets and shore. A century earlier, Baltimore town commissioners in February 1768 had devised a system of draining water from the main street, Baltimore Street. The record specifically states the importance of directing it to the tidewater. The commissioners decided to make a small ditch or canal on both sides of the street, the one on the south (or harbor side) to lead to Gay Street and thence down Gay Street through a canal to the waterside. The north side required a more elaborate route by way of the Jones Falls. Later Ann Street, running downhill to the harbor, was given a concave shape with a one-foot rise to create a storm drain, carrying trash as well as sudden runoffs of water into the Inner Harbor.[40] Later, however, the records of the city of Baltimore between 1782 and 1797 include a resolution in June 1795 "that Daniel Grant be notified to remove a Nuisance on Public Alley, which is occasioned by his necessary."[41] Other nuisances strewed the alleys—and the main streets. Even as late as 1900 stepping-stones remained in use to aid pedestrians in crossing alleys clean shod above the open drains.

Agricultural and industrial wastes also were disposed of in streams and tidewater. Plenty of running water made this flushing easy and effective. But even as early as the seventeenth century, town fathers took care to isolate certain trades from a town's center. In laying out Annapolis, for example, Governor Nicholson imposed an elaborate baroque plan placing centers of church and of government on hilltops and creating separate districts, quite distant, for "Brewhouses, Bakehouses, Dyers, Salt, Soap, and Sugar-Boilers, Chandlers, Hatmakers, Slaughterhouses, some sort of Fish-mongers, etc."[42]

No account of waste disposal can end without a note on the widespread coal mining. In mountainous regions bordering tributaries of the Bay, wastes from this vast industry poured into the network of streams. For example, silt and acid from anthracite coal were dumped into the Susquehanna River's North Branch in Wyoming Valley, Pennsylvania, and bituminous coal wastes went into the West Branch.[43]

Traders and Watermen

The Chesapeake is the Mediterranean Sea of America.
—Late–nineteenth-century publicity

For settlers faced with malnutrition and starvation, the Bay contained godsent foodstuff, and shellfish and finfish were plentiful. This abundance of fish was attested to often in writing; for example, George Alsop, a seventeenth-century observer, wrote:

> As for fish, which dwell in the watery tenements of the deep, here in Maryland is a large sufficiency, and plenty of almost all sorts of Fishes, which live and inhabit within her several Rivers and Creeks, far beyond the apprehending or crediting of those that never saw the same, which with very ease is catched, to the great refreshment of the Inhabitants of the Province.[1]

John Smith noted that the male Indians "bestowe their times in fishing, hunting, wars, and such manlike exercises."[2] And Smith's contemporary in Virginia, George Percy, wrote of the very large oysters roasted by Indians being "delicate in taste."[3] Later the flavor of the seafood also came in for comment: on October 13, 1765, William Gregory noted in his journal, "I put up at Midton's [Annapolis]. Fine oysters to be had." The next day he wrote, "I set out for Joppa about 12 o'clock and arrived here just 8 o'clock. Pretty large vessels coming up to this place. Good oysters. . . . Gunpowder river runs past here."[4]

THE BAY AS A FISHERY

The colonial governments in both Maryland and Virginia received from the Crown the right to dispose of the tidelands subject to public rights of fishing.[5] And exercise their right of fishery the public did.

Even in the first century of colonization some waters were over-fished. In 1678 the Middlesex County court in Virginia acted to con-serve the county's fish, by banning nighttime fishing with fish har-poon irons or similar devices. Certain residents had overfished, in the words of the complaint, "to the Great hurte and greivance of most of the Inhabitants of this County."[6] A century later, in 1796, the Mary-land Assembly passed an act for the preservation of a breed of fish in the Patuxent River: there should be no whipping or beating the water between February 1 and June 1.[7]

Following American independence, the inhabitants of Maryland and Virginia succeeded to all rights previously held by the Crown and Parliament. These rights were to be exercised by the General As-sembly in Maryland and the legislature in Virginia. The Virginia con-stitution restricted this legislative power by prohibiting sale or lease of "natural oyster beds."[8]

Not long after, disputes arose between Maryland and Virginia over fishing rights. The Potomac River served as the major boundary between the states, and Maryland claimed, by virtue of its charter, ex-clusive rights therein. Virginia was taking advantage of its sover-eignty over the capes of the Bay to charge tolls for vessels carrying goods to and from Maryland. In the Compact of 1785 Virginia agreed to give vessels bound for Maryland free passage in return for an agreement by Maryland that the right of fishery in the Potomac was to be equally enjoyed by citizens of both states.[9]

In the Potomac then, the fishery was governed by "mutual con-sent and approbation of both sides." Elsewhere in the tidewater the states went their respective ways. Disposing of the tidelands, subject to public rights of fishery, was the way local governments encour-aged commerce. In 1745, for example, the Maryland General Assem-bly decided that "all improvements, of what kind so ever, either wharfs, houses or other buildings, that have or shall be made out of the water, or where it usually flows shall [as an encouragement to such improvers] be forever deemed the right, title and inheritance of such improver."[10] Then, in 1759, wharves one thousand feet long were built in Baltimore by John Smith, William Buchanan, and William Spear. Without this lengthy wharfage to deep water, Baltimore would not have grown as a port.[11]

From the beginning, fishermen put pressure on legislatures to protect the supply of fish by keeping waters open for the passage of fish. Among these regulations, those in *Acts of Assembly Now in Force in the Colony of Virginia for 1769* ordered owners of mills, hedges, and stone stops to make openings for passage of fish. The courts also ruled against obstructing rivers and streams with milldams, weirs, and hedges. Since these regulations and rulings of course aided navigation as well, much of the wording was general. But many specifically called the obstruction (usually a mill) deleterious to fish.

Examples of the conflict between fishermen and mill owners would fill a book. In Virginia, for example, the legislature passed an act in 1748 that authorized clearing rivers where the passage of fish was obstructed,[12] and it went on in 1759 to order owners of mills on the Rapidan to make slopes for the passage of fish because the inhabitants of Culpeper and Orange counties had complained of obstruction. Within the decade, other acts forced mill owners on the Rivanna, Hedgeman, and Meherrin rivers also to make slopes.[13] Building mills on the Rockfish River below the forks was prohibited, and one Allan Howard had to tear down his mill in 1761 because it was entirely obstructing the passage of fish.[14] The Maryland legislature also enacted legislation, usually about a specific tributary: the Patuxent in 1802, the Monocacy in 1806, the Susquehanna in 1813, and the Pocomoke in 1862.[15]

In these official actions, governmental officials responded to, and evolved a way of dealing with, conflicts over different uses of tributary waters. Back in the seventeenth century the council and burgesses of Virginia apprised the governor of the "great mischiefs and inconveniences" that killing whales in the Chesapeake "accrew to the inhabitants." The reason: By these killings "great quantities of fish are poisoned and destroyed and the rivers also made noisome and offensive."[16] Fisheries were subject to regulations about polluting also. In 1810 the Maryland Assembly ordered them to remove the pickle used in their operations, to clean their slips, and to remove all fish from shores within ten days of the end of the season.[17]

But fish quantity was the goal. One fishery owner, a Virginian named John F. Mercer, boasted in 1797 that he "contrived to land 20,000 [herring] a day." And to his neighbor on the Potomac near Maryland Point, Richard Sprigg, he wrote, "If I had your lease [as well], I could make a fortune."[18] Anne Royall, a Maryland journalist, traveled in the early nineteenth century around the United States and reported on the large fisheries for herring (and shad) at the mouth of the Susquehanna: the fish were taken in large nets "from 108 to 200 fathoms in length spread across the river by boats. The ten fisheries employed fourteen to fifteen men each at fifteen dollars per month, with their provisions, to catch, cure, and pack the herrings in barrels." The author added: "These fishermen make a wretched appearance, they certainly bring up the rear of the human race. They were scarcely covered with clothes, were mostly drunk, and had the looks of the veriest sots upon earth."[19]

Like finfishermen, shellfishermen took all they could—an enormous volume in the late nineteenth century—and jealously guarded their territories. 'Twas ever thus, as an anecdote will illustrate. The enthusiasm for oysters led a group of five seventeenth-century Virginians to break the Sabbath. When discovered, they had a canoeful of oysters. But they escaped punishment, and presumably kept the oysters, because the sick wife of one of them had a craving for oysters, and on Saturday the wind had prevented her husband from going out and tonging them.[20]

Colonists began making a living by oystering early. And by the end of the eighteenth century they had organized an industry. They had borrowed the Indians' log canoe and improved on it for transporting oysters from the rocks (the natural oyster beds). Likewise, they had appropriated the Indians' rake, joined two of them with a hinge, and thus invented the oyster tong that continues in use in the twentieth century.

By 1875 these watermen (as they were, and are still, called) were harvesting millions of bushels of oysters a year. But the supply did not last: "It is significant to note that while the population of the United States between 1880 and 1930, increased by 144.8 percent, the per capita oyster consumption declined by 77.3 percent." In this period there was an 81.6-percent decline in annual catch. In 1880 Maryland had 47.5 percent of the Atlantic and Gulf production, and by 1930 it had only 15.4 percent.[21]

When necessary, watermen looked to the courts to protect their rights in God's oysters. At the time of the Civil War in *Phipps v. Maryland*,[22] a defendant was accused of taking oysters from a private bed. He unsuccessfully defended himself on the grounds that there was a common-law right of public fishery. Apparently, however, such rights do not extend across state lines. A decade after the Civil War in *McCready v. Virginia*,[23] the U.S. Supreme Court held that the Virginia legislature could lawfully exclude a Maryland resident from taking oysters in Virginia waters.

Watermen also looked to the legislatures. A great volume of legislation in the 1800s concerned fish and oysters, but mostly oysters. State legislatures, particularly Maryland's, tried to control oystering and to maintain the supply. One student of the subject repeats the legend that, ever since the law against dredging in 1820, Maryland legislators have enacted more laws on oysters and oystering than on all other topics.[24] For example, the Maryland Assembly, responding to the concern of oystermen, in 1834 banned improper seines in use that were covering oyster beds with mud and thus diminishing the harvest.[25] In 1860 the Assembly set seasons for oystering.[26] It also imposed fines for throwing oyster shells in the water. In Virginia laws had been on the books since 1691 prohibiting the "casting of Ballast into Rivers and Creeks" and requiring ballast to be unloaded on land above the high-water marker.[27] In 1748 the Virginia legislature authorized overseers to direct the delivery of ballast to the shores.[28] No doubt such laws were passed to aid navigation as well as to protect fisheries.

NAVIGATION

From Brother Carrera in 1572 on, no one overlooked profit from trade. Many factors contributed to successful commerce in what a nineteenth-century puff called "the Mediterranean Sea of America." For one, even if the Chesapeake doesn't exactly match the Mediterranean climate, it nevertheless remains open year-round. There is little fog. And, as an observer said of Richmond when that city became the capital of Virginia in 1779, it is "more safe and central than any other town on navigable water."[29]

Another advantage derives from the rush of fresh water that spills into tributaries of the Bay from the piedmont falls. In the past, that fresh water was used to kill what was called the termite of the sea, the shipworm, or teredo, enemy of wooden ships. The waters of the Jones Falls in Baltimore's Inner Harbor cleansed many a sailing ship.

Profitably and significantly, the sprawling tidal estuary permitted ships to sail far inland. This realization occurred to William Byrd III in laying the foundations of both Richmond and Petersburg in the

1730s. He wrote that "the uppermost landing of the James and Appomatox rivers . . . are naturally intended for marts where the traffic of the outer inhabitants must center. Thus we did not build castles only, but also cities in the air."[30]

Merchants especially worked to promote the navigability of these rivers. They sought the help of both courts and legislatures to keep ships moving smoothly wherever trade beckoned.

From the beginning, Virginia lawmakers focused whatever attention they gave to water on its usefulness as a road. This outlook was based on economics, not twentieth-century concerns with pollution or siltation.

Virginia enacted numerous laws, to deal with disposal of felled trees in waterways, that managed improvements to navigation on

By the end of the eighteenth century Hampton Roads, Virginia, was a prosperous trading center with several harbors, among them Norfolk. This sketch and watercolor drawing comes from the journal of Benjamin Henry Latrobe when he came from England as a young, unknown architect in 1796. Courtesy the Library of Congress.

every major river. When in 1679 a grand assembly met at St. James City, it passed a law authorizing the clearing of rivers near their head-waters of logs and trees. Furthermore, the counties were left to decide when and how to conduct improvements of the rivers to facilitate the passage of ships.[31] Then in 1680 the legislature authorized appointment of county surveyors to clear waterways. Cited was the loss of boats because "logs, trees, etc." were not cleared. A fine of five hundred pounds of tobacco was imposed for obstructing navigation.[32]

Again in 1722 the legislature passed an act for the "More Effectual Clearing of Rivers and Creeks." In the preamble to the act, the legislature declared that "many of the rivers and creeks of this colony are stopped and choked up by the fall of trees, stumps, and rubbish . . . and hedges are made across the same whereby the passage of . . . vessels is hindered and obstructed . . . to the great damage of the inhabitants . . . [and the] hindrance of their trade and commerce." The counties were then authorized to contract to clear the streams.[33] A later act, in 1726, especially noted that the debris thrown into the water not only obstructed navigation but also caused bridges to break apart and get washed away.[34]

Since tobacco shipping dominated trade in the first two centuries, certain acts specifically referred to the more convenient transport of tobacco. For example, one hundred pounds, appropriated at the time, was to be spent to clear the Fluvanna River of rocks.[35] In Maryland also, the legislature and courts helped keep channels open. In *Harrison v. Sterett*, decided in 1774, the provincial court of Maryland granted relief to one waterfront lot owner against a neighbor who was constructing a wharf which would have restricted the former's water access.[36] And in 1768 the Maryland Assembly had passed an "Act to Prevent Obstructions in the Patomack and Monocacy Rivers" caused by fish dams.[37]

Townspeople insisted that leaders manage shipping in the harbors. Ports around the Bay revived the ancient English post of port warden for the purpose. The warden's duties included conducting surveys of channels, keeping them clear and clean, and seeing to the removal of obstructions and annoyances. The Maryland Act Appointing Wardens for Baltimore Town in 1783 (just before the incorporation of the city) permitted the wardens to pass regulations and ordinances preventing injury to the harbor from wharf construction, discharge of ballast, or any other cause.[38] They were specifically asked to approve all wharves that diverted the channel or obstructed navigation.

In the Compact of 1785, Maryland and Virginia agreed that their respective citizens could share navigation rights over the Potomac River, and that Virginia would give ships bound for Maryland free access to the Bay. An act of the Maryland Assembly in 1791 authorized the commissioners of the city of Washington to enact regulations governing the discharge of ballast from ships in the Potomac above the lower line of the District of Columbia and Georgetown and in the eastern branch.[39]

Ballast discharging bothered Baltimoreans too. But, more important, a persistent concern of Baltimore's merchants was the shallowness of the Inner Harbor, where the wharves were both numerous and long. Wharves were numerous because the chief merchants ruled from there; they were long because they had to reach out to the seagoing ships. The shallowness was being worsened by the deposit of sediment washed down the Jones Falls into the narrow basin. An act of the Maryland Assembly in 1791 authorized the harbor's deep-

Running the Rapids of New River, Virginia. This wood engraving after a drawing by W. L. Sheppard, in *Harper's Weekly*, February 21, 1874, shows how rivers were used to bring products down to steamboat landings, and then to markets. Courtesy the Library of Congress.

ening.[40] And after the War of 1812, leading Baltimore politician and merchant Samuel Smith persuaded the federal government that the ships sunk in the channel as defense against the British attack on the city in September 1814 had to be removed at the federal government's expense. The funds obtained not only raised the ships but also deepened the channel.

That improvement came at just the right time. The first steamboat on the Bay, the *Chesapeake*, had been built in Baltimore in 1813. Soon the nineteenth-century steam revolution filled the waters with trade. Just within the Bay region a number of steamship lines connected major ports as well as small landings like Lodge Landing on the South Yeocomico River, Virginia. For one hundred years and more, everyone knew the routes and names of such lines as Weems and of steamships like the *Lancaster* and the *Maggie*—and others with names of females best known to shipowners. Clearly, the Bay world shrank with the first steamboat.

Because of their speed and certain schedules, steamboats replaced sailing packets on the route from Baltimore north to stage connections with Philadelphia. After the Chesapeake and Delaware Canal opened in 1829, boats linked the two cities. The *Chesapeake* of course was a harbinger—and not just for Bay routes. Soon coastwise steamers arrived, and then overseas ships. Their histories fill volumes. By 1972, oil tankers created floating islands as they waited to enter the Baltimore Channel.

From those days of sail to today with our huge tankers and containerships, Bay ports flourished. How great a volume of trade these ports brought can be told from the high trade rankings of Baltimore and Hampton Roads. These two ports hogged the waterborne trade after the Civil War. Before the war, however, docks upstream served the hinterland well. Courts and legislatures saw to maintaining channels to upper estuary towns like Richmond and Georgetown.[41] Virginia and Maryland legislators were forever opening and extending navigation beyond the settled districts. In 1800, Maryland acted to remove bars, shoals, and other obstructions from the Anacostia River near Bladensburg.[42] But the act did little good because the siltingup continued, and Bladensburg became (and remains) a backwater isolated by silt. The state in 1814 created the Potomac Company to extend and improve navigation on the Potomac,[43] and in 1817 authorized the use of lottery revenues to improve navigation.[44]

A typical broadside showing how advertisements attracted tourists to the newest in water transport in the 1860s. For six dollars Washingtonians could steam down the Potomac to Norfolk, Virginia, and back, meals included. Courtesy the Library of Congress.

An unfortunate effect of settlement on the Bay's shores was erosion. Early settlers wrote of soil erosion and pollution, and late seventeenth century visitors noted erosion and the muddiness of freshets.[45] By 1800 silting had destroyed the channels leading to prominent tobacco ports at Joppatowne, Port Tobacco, and Upper Marlboro in Maryland. On the Gunpowder River, Joppatowne once took an eight-foot draft. But just between 1848 and 1897, seventy-nine hundred cubic yards of sediment were deposited in the upper part of the Gunpowder River estuary.

To blame was careless use.[46] In 1753 the Maryland Assembly passed an "Act to Prevent Injuring the Navigation of Baltimore-town and to the Inspection House at Elk Ridge Landing on the Patapsco." The legislators noted that people were digging into banks of that river, causing large quantities of earth and sand to be washed into the water. So they also made it unlawful to put earth, sand, etc., into navigable branches unless well secured.[47]

In 1783 the Maryland Assembly gave the port wardens power to enact regulations to prevent injury to the harbor from construction land, earth, or soil contiguous to the basin or harbor that might fill it

Town or Location	Founded	River or Creek	Approximate Time Sedimentation Recorded	Amount of Downstream Migration of Head Navigation (Miles)	Approximate Reduction In Depth (Feet)	Years
Bladensburg, Md.	1742	Anacostia	1875	2	3	1875-1937
Piscataway, Md.	1634	Piscataway	1807	1	3	1863-1945
Georgetown, Washington, D.C.	1751	Potomac	1804	20 (No Dredge)	9.25	1783-1837
Mt. Vernon	1752	Potomac	1793	—	1 to 4	1863-1904
Dumfries, Va.	1748	Quantico	1787	1.7	4	1796-1905
Iron Factory	1734	Neabsco	—	0.75	—	1734-1872
Port Tobacco	1658	Port Tobacco	1700	1	6	1800-1882
Upper Marlboro	1706	Patuxent	1733	8	7	1859-1944
Elk Ridge near Baltimore	1650	Patapsco	Before 1898	7	15 at Hanover St.	1845-1924
Joppa Town	1707	Gunpowder	1750	2.5	10	1750-1897

Professor Gordon Wolman of Johns Hopkins University compiled this table for his study for the Governor's Conference on Chesapeake Bay, 1967.

up or obstruct it.[48] A century later the assembly prohibited disposal of ballast or earth in the Bay itself.[49]

The main problem centered in Baltimore, where citizens worried about silting from at least the 1780s on. Lowering and grading Hughes Street at the foot of Federal Hill caused continuous erosion into the harbor.[50] As construction spread northward along Jones Falls, the mouth of that stream built up alluvium in shoals.

From 1800 on, regular dredging was done in the Potomac at Georgetown, Washington, and Alexandria. By 1863 soil pollution posed a general problem along the Potomac.[51] Today 25 to 30 percen' of the estimated one million tons of sediment entering the Potomac yearly comes from construction sites in the District of Columbia area. National Airport is built on dredged channel sediment.[52]

The mid–nineteenth century produced a clutch of cases concerning efforts to improve navigation. In *Garitee v. Baltimore*, the city deposited dredge spoil in front of the plaintiff's property; the court granted the plaintiff relief.[53] In *Baltimore & Ohio R. R. v. Chase*, a dispute arose between neighbors as to access rights; the court preferred

Richmond from the Hill above the Waterworks. Richmond, Virginia's capital, has produced a body of law governing uses of Chesapeake waters. This view in 1834 shows the James River and its canal. Through canals like this one, products of the piedmont reached ports on the fall line such as Baltimore and Richmond. These idealized views made Chesapeake towns seem both handsome and prosperous. Courtesy I. N. Phelps Stokes Collection, the New York Public Library.

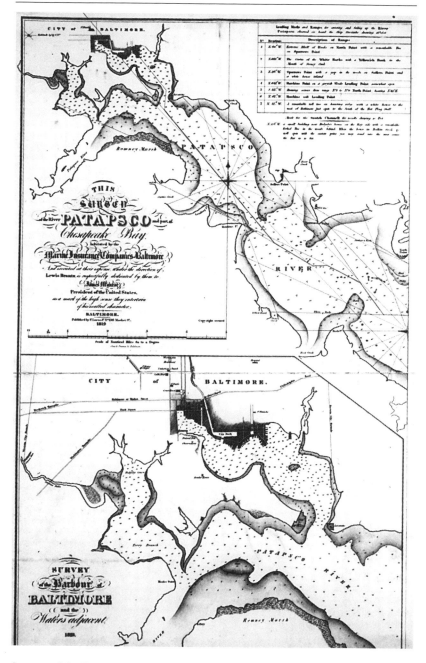

Survey of the Patapsco River and part of the Chesapeake Bay by Lewis Brantz
in 1819. Published in Baltimore by Fielding Lucas, Jr., the charts provided

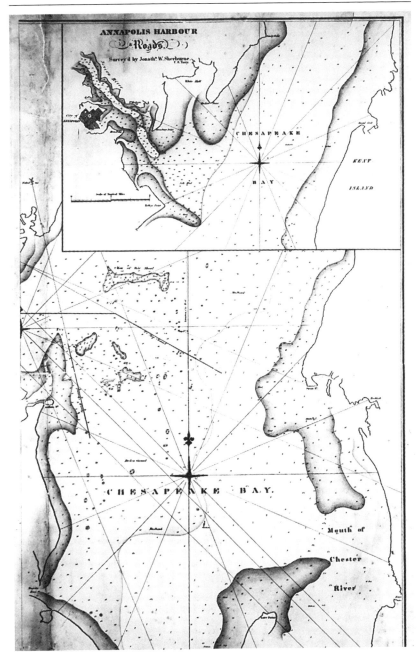

information that was essential to the growth of the port of Baltimore in the nineteenth and twentieth centuries. Courtesy Maryland Historical Society.

the rights of the railroad, which had already constructed an improvement to navigation.[54] In *Norfolk v. Cooke*, the city was granted relief, which precluded a private party from filling in the tidal basin.[55]

In 1890 the federal government also began regulating navigation. The U.S. Army Corps of Engineers established lines in various harbors beyond which piers or deposits could not legally be placed. The corps also worked at dredging and constructing improvements, like the channel to Baltimore. More will be said about their projects in the discussion of the proposal to enlarge the Chesapeake and Delaware Canal. This canal, opened in 1829, became the most enduring of the series of canals built in that era. The Chesapeake and Potomac Canal and the ones on the Susquehanna River lost out to railroads in the rivalry for freight and passengers. We should add that Pennsylvanians in particular seem to have been keen to develop navigation. Investments by the state as well as private entrepreneurs went to building shipyards, steamboats, and hundreds of miles of canals on the Susquehanna and tributary rivers. Some Pennsylvanians even advocated blasting and dredging to make a deep channel in the shallow, obstacle-ridden Susquehanna River.

By the end of the first three hundred years of the white man's use of the Bay, another set of improvements came under the eye of the federal government: lighthouses. The U.S. Coast Guard took a role only long after the individual states had erected most of the seventy-four that were built. The states began building them when steamboats arrived. In the early times, lighthouses stood like the bobbins of a knitting factory, guiding the skein of steamship lines that stretched all over the Bay.

The lights also guided fishermen as they "traveled the dark." A number of these lights still function, and four of them were still manned as late as 1970. Others, abandoned, stand recycled as houses, or in ruins, or else they do not stand at all. Once the Bay's shores held more screwpile-type lighthouses than any other place. One curious historical fact: though the Chesapeake had the greatest volume of shipping in eighteenth-century America, it was the last important region to get what one man called "noted landmarks to guide the doubting mariner."[56]

Lighthouses, however, made only slight territories for the federal presence. More significant, the seat of the federal government itself faced the Potomac. Congress had also taken over more and more of Annapolis for the U.S. Naval Academy since the school moved there in 1847. Then too, the Navy Department created a major base for its ships in Hampton Roads at Gosport after the Civil War. In 1917 the deep-water harbor there became the greatest naval base in the country. This record size points up the shift from fish to ships in the first three hundred years of the Bay's history. This naval base also has significance because it illustrates the twentieth-century dominance by federal presence. But much more of that story will follow in succeeding chapters.

6
Land of Pleasant Living

Chesapeake Bay, land of pleasant living.
—Mid-twentieth century Baltimore beer commercial

W e must remember that boats were sailed for pleasure as well as commerce. For over two centuries men and women skimmed over the surface of Chesapeake waters, merry and free. Joseph Seth, for example, recalled in 1926 the sailing canoes of the 1870s: they were swift sailors—never surpassed, he wrote, for beauty of line. "The days spent in our 'Flying Fish' were some of the happiest of my life." He noted that even the small canoes were equipped with sails, that almost every family possessed one, and that this kind of boat had played an important part in all regattas since colonial times. Of sailing parties by moonlight, Seth remembered, "The union of the beauty of the night and the graceful flying canoe were the essence of poetry and romance."[1]

Since the seventeenth century, private boats like the *Flying Fish* have been sailing pleasurably across uncrowded ways. In 1689, for example, Mayor Robert Sewall and his ketch *Susanna* provided a footnote to recreation on the Bay. Then in 1760 we have record of an organized regatta. Later in that century Edward Lloyd of Wye on the Eastern Shore wrote his London agent: "Be pleased to send me a complete set of American colors, for a pleasure-boat of about 60 tons burthen and six brass guns to act in such manner as to give the greatest report, with the letters E.Ll. thereon."[2]

In 1825, the Baltimore *American* advertised pleasure boats for hire at Fells Point, Baltimore. No wonder artists painted idyllic waterscapes with backdrops of domes and steeples that make the city look like Constantinople. Even back then, the Inner Harbor held more than just commerce.

Around 1800, lower down the Bay, the schooner *Dolphin* carried "on board a number of Gentlemen on a party of pleasure." The reporter added that, equipped with a seine and a variety of fishing tackle, they were "abundantly supplied with fish . . . and to crown their felicity, they found no vessel in the bay able to sail with them [that is, capable of outsailing them]."[3]

Such yacht racing was not for everyone. But luckily, the masses could also make "parties of pleasure" on the Bay, once excursion steamboats began plying the waters after the Civil War. In summer, people flocked aboard to cool off. Some passengers spent the day at resorts like Chesapeake Beach, in Calvert County, Maryland, or on the Eastern Shore at Tolchester or Betterton. In its brochure, Chesapeake Beach offered "Finest Salt Water Bathing," fishing, crabbing,

The clean, white excursion steamboat contrasts dramatically with the smoke-darkened horizon of Baltimore. Belching smoke on the left is the power plant that ran streetcars for years. In 1905 when this photograph was taken, excursionists returning from Betterton across the Bay always commented on the fact that they knew they were within five miles of Baltimore by the putrid harbor smells. Courtesy the Photography Collections, Albin O. Kuhn Library and Gallery, University of Maryland, Baltimore County.

yachting, and "attractions not offered by any other Resort this side of Atlantic City."

In the 1820s, with the coming of the steamboat, Virginia excursionists steamed to bayside resorts too. Some people went there with special purpose, to sustain or regain health. Health seekers basked in the salubrious air at Old Point in Hampton Roads, where in 1821 a hotel opened, appropriately called Hygeia. But convalescents merely opened the way. Soon they were outnumbered by "public men, weary of their cares, army and navy officers furloughed or retired, and the gay daughters of Virginia."[4] As early as 1854, Ocean View, Virginia, attracted invalids because it offered quiet and recreation. Any invalid or convalescent who was up to exercising could have walked on the beach, ridden horseback, hunted ducks, or fished.

Through all the later decades of the nineteenth century, recreation abounded for both invalids and the careworn. Who wouldn't have found pleasure in this region of clean estuaries and rich marshes? A sky full of ducks and geese brought hunters. By 1814, we know, hunters were using "wooden figures cut and painted to represent ducks."[5] Then to meet another need, clever breeders mated Newfoundlands with water spaniels to produce the Chesapeake Bay retriever.[6] Down the years human ingenuity in devising decoys, dogs, and blinds balanced human laws such as bag limits and set seasons—all to the furtherance of man's pursuit of a duck.

Hunting stories take second place in glory only to fishing tales. Nostalgia infuses both. Remembering 1850, Joseph B. Seth wrote:

> My mother was a great lover of fishing. . . . It was her almost daily custom in August and September to paddle out on the water in her special rowboat about 4 o'clock in the afternoon, stopping to throw her lines not more than twenty-five yards from the foot of the lawn. Here she would fish for two hours. Her luck was proverbial, as she rarely failed to bring home a good catch. These fish were often used for our supper, literally almost going into the pan alive and kicking. Good fish eaten under these conditions made delectable food.[7]

The same enthusiasm was demonstrated for crabs, though it developed later. In 1836 a Philadelphia doctor reported that "soft crabs are, with great propriety, regarded as an exquisite treat by those who are fond of such eating," although he noted that many people wouldn't touch either lobsters or crabs. He added, "There are few who taste of the soft crabs without being willing to recur to them."

This Philadelphia physician, Dr. John D. Godman, called soft-shell crabs "an article of luxury." He continued: "They are scarcely known north of the Chesapeake, though there is nothing to prevent them from being used to considerable extent in Philadelphia, especially since the opening of the Chesapeake and Delaware Canal [in 1829]."

He provided other information that may startle twentieth-century crab eaters:

> The price at which they are sold is sufficient to awaken all the cupidity of crabbers. Two dollars a dozen is by no means an uncommon price for them when the season first comes on. . . . The slaves search for them at night, and then are obliged to kindle a fire of pine-knots on the bow of the boat, which strongly illuminates the surrounding water, and enables them to discover the crabs.[8]

A more modern scholar, L. Eugene Cronin of Chesapeake Biological Laboratory, asserted that crabs positively had been eaten near the Chesapeake since 1855, when crabbing was the basis for a small industry. Crisfield entrepreneurs shipped crabs from the 1870s on. Back in the 1730s William Byrd ate "flat crabs" about as large as a hand.

Pleasure in swimming in Bay waters must have peaked around the turn of the twentieth century, or even later. This recreation did not last. Indeed, the custom of bathing from beaches had begun for the masses only late in the nineteenth century. Earlier of course it would have been hard to keep boys out of the water. Now a dignity far above the Tom Sawyer level of skinny-dipping settled over resorts like Virginia Beach, Virginia, and Betterton, Maryland. There ladies in black stockings and skirted bathing dresses disported on the sand, as they appeared at the opening of the first municipal bathing beach in Baltimore, a stretch of Canton shore just east of Fells Point, in 1893. This spa was fitted out with floats, ropes, and crude cabins for changing. Suits could be rented. So popular was recreating in the rather polluted Patapsco waters that the number of bathers during the second

Fortress Monroe, Old Point Comfort, and Hygeia Hotel, Virginia, in 1861 and 1862. The Key to the South. In the 1840s vacationers and invalids escaped the cities and, in places like the Hygeia Hotel, with its stately columns, found healthy recreation on the Bay. Courtesy the Library of Congress.

season of 1894 was 27,787, with the number varying from twenty-five thousand to forty thousand from 1896 to 1898.[9]

Other escapees from August heat waves danced the evenings away as they cruised on the *Emma Giles* or sister ships. How many people cruised we can learn from the account of an accident. Fourteen hundred excursionists aboard the *Louise* were returning to Baltimore from Tolchester on July 28, 1890. About 8:15 P.M. as they were nearing Seven Foot Knoll, their ship was hit by the *Virginia,* a steamer then on its regular run from Baltimore to Norfolk. Fourteen people were killed.

At that time and just east of this beach, Back River neared the end of its long usefulness for recreation. Sportsmen had loved hunting and fishing there. Then Baltimore City authorities chose 547 acres at the western end of the Eastern Avenue bridge for the construction of a sewage treatment plant.[10] (The significance of this plant for other aspects will be discussed in chapter 7. For our discussion here, however, it meant the end of pleasure for sportsmen at this site.)

Still, other Bay waters and marshes filled the need for a time, though more distant from the metropolis. It is important to recognize that the compact character of rowhouses had always given Baltimoreans easy access to the country. In the early twentieth century, however, the city began to sprawl. One last chance to preserve nearby land and water came in 1904 when the famous Olmsted brothers' firm of planners came to town, hired by the Municipal Arts Society of Baltimore, a private group. They came fresh from having created a chain of parks for Bostonians, including Riverside near the mouth of the Charles River. In Baltimore, the planners expressed dismay that no public park opened the Bay to citizens.

They found the perfect spot for just such a water park only seven miles south of the center of town. They chose Curtis Creek and its tributary Furnace Creek. There, the team wrote, eight hundred acres of land, which could enclose eight hundred acres of water, could create an area ideal for small boating, picnicking, and swimming and other water sports. This new park would give Baltimoreans access to Curtis Bay, to broad and quiet tidal estuaries, to "scenery which forms the most beautiful and characteristic landscapes of all the country lying along Chesapeake Bay."[11]

Nothing came of this proposal, so Bay lovers lost an amenity; the Olmsteds' genius lost a monument; and Baltimoreans lost a nearby escape from crowded, hot streets on Chesapeake waters.

Certainly town dwellers had to escape then just as they do now: Tidewater towns, it would seem, have never been agreeable places to live. Hampton, Virginia, for example, stands as the oldest settlement of English America in continuous existence, but its marshes and mudbanks early made it an unhealthy spot to live in. A visitor in 1796, Isaac Weld, reported that Hampton was "a dirty disagreeable place always infested by a shocking stench from a muddy shore when the tide is out."[12]

Again, about Norfolk, Virginia, William Byrd II wrote, "With all these conveniences [for trade], it lies under the two great disadvantages that most of the towns in Holland do by having neither good air nor good water."[13] Other locations around the Bay, happily, did possess the air and water deemed healthy by eighteenth-century colonists. Both Richmond and Baltimore benefited from sites on the fall line, with hills and fast running streams.

Norfolk suffered another drawback. Visiting there in 1794, Mederic Moreau de St. Mery complained about the new houses springing up in the direction of the Elizabeth River and of the new wharves built in haphazard arrangement, "put up solely for the convenience of the owner and built without any general plan, [which] inconsiderately shut off the view of the river."[14] Trust a Frenchman's eye for the esthetic to notice what was wrong with Bay building: waterfronts filled with warehouses. Men like Byrd exploited the waters without a thought to esthetics (or water quality).

Away from towns, this French visitor might not have found fault. Being domiciled above, near, and on the water satisfied many a man or woman's esthetic sense. Recollecting Eastern Shore life in the mid–nineteenth century, Joseph B. Seth (born in 1846) wrote:

No pleasanter memories arise in my mind than those associated with the beautiful sheet of water upon which our home was situated. The water was salty, it being tributary to Chesapeake Bay, and flowed in and out twice a day. It was not a rapid tide, but amply sufficient to keep the water pure.[15]

About the time Seth's great-grandfather surveyed this beautiful property in 1780, George Washington was appreciating the location of his house on another Chesapeake site: "No estate in United America is more pleasantly situated" than his Mount Vernon.[16] Contemporary painters always showed the mansion crowning its plateau above the wide Potomac with ships of trade sailing past, as on page 29. Similarly placed above tidal waters, the state house and governor's mansion in Annapolis and the capitol in Richmond invited people on shipboard to admire how architecture ordered the wilderness (see illustration on page 61).

Lacking such grand vistas of water, English kings and dukes constructed artificial canals and lakes like those at Hampton Court and Blenheim. More fortunate, Chesapeake planters created grander prospects near Annapolis, such as Java (on the Rhode River) and Tulip Hill (on the West River). Builders at Mulberry Fields in southern Maryland carefully framed the water. To make the river, a mile away, seem

Popular recreation at waterside resorts around the Bay included swimming, sunning, and sometimes an opportunity to ride on a Ferris wheel or a merry-go-round. Pictured here is a typical beach at Arlington, Virginia, in 1925. Many beaches were later closed to swimming. Courtesy the Library of Congress.

nearer, they created an illusion in perspective. They planted the avenue of trees flanking the view to fan out from the house to the water.

The special appeal of views of the Chesapeake is shown in these real estate advertisements from the *Baltimore Gazette and Daily Advertiser:*

> For Sale or Rent A very valuable Farm in Harford County, beautifully situated, commanding a fine view of the Chesapeake Bay and Gunpowder River . . . a never failing spring of excellent water at a convenient distance, and a well within a few steps of the door . . . 500 acres, 17 miles from Baltimore . . . Col. Edward A. Howard [Tuesday evening, August 26, 1828]

To rent or lease for a term of years. The Country Residence of the late Alexander H. Boyd, not more than 20 minutes walk from the centre of the city. . . . The view of the River, Bay and surrounding country is not surpassed by any residence near the city. [Friday evening, February 29, 1828]

No matter how agreeable the amenity of a view, colonists had a practical reason for choosing the site—health. To them, a healthy site meant a hilltop with good air. The Lee family's mansion, Stratford, for example, rises some distance from the Potomac River. The builders, however, did erect two platforms within groups of huge chimneys, vantage points from which we can still follow the band of water to the horizons. Though the unusual chimneys resemble designs by Sir John Vanbrugh for Claremont in England,[17] the view belongs to the Chesapeake.

Occasionally, though, vistas and health gave way to placing house and office near warehouses and shipping. Sited close to the James River, Westover (1730) was headquarters for William Byrd's 200-square-mile empire. Byrd and the other planters tell us a lot about attitudes towards governing Chesapeake waters. These eighteenth-century men had to accept change. And, in their isolation, they had to live with independence from government control. The uncluttered landscape and waters of course imposed their own demands for adaptation. Even the architecture changed in the new climate. Curiously, no single English house served as source for Virginia's mansions, though prototypes may include Coleshill, Berkshire (designed probably by Roger Pratt about 1650), and Thorpe Hall, Norfolk (Peter Mills, architect, 1654–1656).[18]

Today we envy planters like the Byrds and the Lees. We wish we had their amenities—uncrowded space, clean air and water, the slow pace of life. Few of us in the twentieth century can spend the afternoon the way this Virginian and his wife did in 1759 and 1760: "September 20: Fine weather. Went in the afternoon and drew the seine. Had very agreeable diversion and got great plenty of fine fish." His diary records other good times:

September 26: Went with my wife in the evening to draw the seine. Got sixty green fish [black bass] and a few other sorts. . . . October 6: Went with my wife to see the seine drawn.

We dined very agreeably on a point on fish and oysters. . . . January 22, 1760: Bought about 70 gallons of rum. Got fine oysters here. . . . February 9: Went with my wife and Mr. Crisell to draw the seine. We met in Eyck's Creek a school of rock, –brought up 260. Some very large; the finest haul I ever saw. Sent many of them to our neighbors. . . . February 22: Drew the seine and got 125 fine rock and some shad.[19]

7

An Immense Protein Factory

Baltimore lay very near the immense protein factory of Chesapeake Bay, and out of the Bay it ate divinely.

—H. L. Mencken, 1940

We have discussed the development of the Bay region over its first three hundred years, and described the origins of various uses of the Bay and its resources. By 1900 the dominant Bay uses of waste disposal, navigation, and fisheries were well established, and there was considerable recreational activity as well. Conflicts among these uses had been few. But four important forces had developed which would directly influence the Bay. Foremost was the development of the cities of Hampton Roads, Richmond, Washington, and Baltimore. With them came the need for water supply and municipal sanitation. Concurrently developing was the large-scale use of the Bay as a waste placement sink. This use was related both to the growth of cities and their water supplies, and to the growth of industry. Deepwater navigation by merchant ships was also on the rise. Last in economic importance, but perhaps first in terms of political attention, was the growth in commercial fishing, most particularly the oyster fleet.

The primary market for commercial fishing on the Bay prior to the nineteenth century was local and attracted relatively little political attention. By the early part of the nineteenth century, New England oystermen were making raids on the abundant oyster bars of the Chesapeake. Not only were the New Englanders taking oysters to northern markets centered in New York and Philadelphia, but they were also using Bay seed oysters to restock their bars, which had been stripped by earlier generations of New England fishermen. Both Virginia and Maryland moved to prevent out-of-staters from capitalizing on the Bay's resources, and there gradually developed a robust Bay oyster industry, which peaked in the 1880s.[1] Despite ex-

Log canoes provide some of the best pleasure sailing in the country. The well-known photographer A. Aubrey Bodine captured a scene on the Miles River near St. Michaels around 1935. The lead boat is *Magic*, built in 1894 by Charles Tarr in St. Michaels and owned by the Wilson family of Talbot County since the 1920s. Courtesy the A. Aubrey Bodine Collection, Peale Museum.

clusionary legislation, much of the impetus for this development still came from New Englanders, who financed many of the largest harvesting, processing, and shipping operations.

Development of the oyster industry has been a dominant factor in the evolution of Bay water-quality policy and politics for nearly a century: Virtually all significant water-quality issues have had the welfare of the oyster and the oysterman as a central concern. It is, therefore, useful to outline the development of the industry in order to explain its central importance.

Oysters in the Chesapeake Bay are nearly ubiquitous except in its upper reaches and tributaries, areas of low salinity. They occur in commercial quantities on beds variously known as bars, reefs, rocks, or knolls. These bars are potentially self-sustaining. Oyster larvae spend a period of time as a free-floating part of the Bay zooplankton but eventually must settle on a hard surface to begin the development of a shell and their growth to maturity. An ideal surface is the shell of another oyster, living or dead. Thus, over time, a favorable site will develop an expanding area and a quantity of oysters built on the shells of oysters that preceded them. In some places these bars grew to within a few feet of the surface, hence the name "reef." It was these natural bars that attracted the New England oystermen and later provided the crop that reportedly totaled over twenty million bushels a year during the 1880s.[2]

Bars built by nature can also be destroyed by natural means. Long-term freshening of the upper Bay and the Potomac have eliminated once-productive beds, leaving vast quantities of shell. The constant shifting of the Bay bottom and erosion of its shoreline have silted over bars that could not grow upward fast enough to keep ahead of the deposits. And natural enemies such as the starfish, the moray eel, various smaller predators, and diseases could, and no doubt did, eliminate bars. But in the nineteenth century, and by general reputation still in the twentieth, the Bay was perhaps the most productive natural oyster ground on earth.

Man also can both destroy and build oyster beds, in the latter case requiring assistance of nature. Destruction can be indirect through the modification of salinity patterns in the Bay or by increasing sedimentation and erosion.

But the most dramatic destruction of oyster beds comes from overfishing. Overfishing made large areas nonproductive during the heyday of the oyster raids of the late nineteenth century. Despite growing concern for the decline of the industry, and a large body of law to prevent it, beds were stripped of their standing crops of mature oysters. In some cases, beds were rendered unfit for future productivity.

On the positive side, man can establish a bed for oyster production, using techniques perfected in France in the early part of the nineteenth century. Oyster shells can be dumped in an otherwise fallow area, and either receive a natural set of spat, or be provided with seed oysters from another area. Through a process of preparing the ground, planting, thinning, and preventing excessive siltation and predators, man can produce oysters in a fashion analogous to land farming. The same techniques can increase the productivity of natural bars.

Government involvement in the oyster industry was, as mentioned, extensive in the nineteenth century. The "oyster question" occupied much of the time of the Maryland and Virginia legislatures. The first task was to prevent outsiders from stripping the bounty of the Bay. The second was to prevent residents from overfishing. The third was to resolve the various disputes between different counties, since most regulation was done on a county-by-county basis. The fourth task was to address differences between hand tongers, who generally worked the more protected and shallow waters of the Bay's tributaries, and the dredgers, who worked the larger bars and were regarded as the chief sources of oyster bar destruction.

Last, but not least, legislatures had to address the question of whether to allow the leasing of Bay bottom for the farming of oysters. Clearly it would not be profitable to go to the expense of developing a bed if someone else could harvest it. Many observers of the great oyster boom believed that the future of the industry lay in private oyster culture. But the oystermen who made their living harvesting naturally grown oysters did not want to be denied that opportunity.

Virginia regulated a compromise. The solution was to establish that all natural bars, as delineated in a late–nineteenth-century survey, were to remain as common property resources, open to harvest by anyone who met residency requirements and who observed laws relating to licensing, gear, and season. At the same time, a method was established to lease fallow bottom to individuals or corporations, thus permitting the development of a substantial private culture as had been developed in Europe, New York, and New England.

In Maryland, the oyster was seen to be a prime economic asset.[3] Since 1884, its fisheries officials had argued that the Bay was a tremendously productive body of water, capable of producing far more oysters than occurred naturally. The institution of oyster farming, following the methods developed in France, could make the yields of the Bay greater than those even of the heyday of the great raid on the natural bars during the 1880s. (Indeed some officials at that time argued that the faster the natural beds were depleted, the sooner the state could turn to the proper management of the resource, allowing for a stable and economically important industry comparable to dry land farming.[4])

For complex and varied reasons, the concept of private oyster farming was never accepted widely in Maryland. Factionalism between watermen from different counties, between dredgers and tongers, and between all watermen and the conservation agencies drained whatever energy there might have otherwise been to get about the business of managing the oyster more efficiently. The only thing the watermen agreed upon was their opposition to a private oyster culture.

Maryland oystermen formed the nucleus of what was to become a powerful political force. Since oysters had to be shucked, canned, packed, and shipped, subsidiary industries employing thousands of workers had sprung up throughout the region.[5] The economic atlas for the state of Maryland at the turn of the century illustrates not only the numbers of persons involved in the industry, but also the fact that they were distributed throughout the tidewater counties.[6] Because each county then had a single senator and also because the tidewater counties outnumbered their nontidal counterparts, political representation for the oyster interests was disproportionately large.[7] By the turn of the twentieth century the Maryland oyster lobby had political clout greatly in excess of the economic importance of the oyster industry.

The commercial harvest of finfish has never gained anything like the size or political significance of the oyster industry in the Bay states. It was a matter of considerable attention in the nineteenth century, however, particularly the shad harvest. Shad runs had apparently long supplied a substantial portion of the food of some of the poorer Bay area residents, and reductions in spring runs had been lamented at least since the 1920s. The numerous statutes relating to the damming and blocking of streams were aimed partly at reducing the obstructions to the spawning of shad. Declines in supply had the effect of driving up prices, a rise that was decried as having an unfortunate impact on the poor.[8]

Declines of finfish through the eastern United States had much to do with the formation of the Federal Fisheries Service. This agency in turn led to the establishment of state fisheries boards in both states soon after the Civil War. Along with the oyster police force also established during this period, these activities constituted the first steps to creating state bureaucracies for the management of Bay resources. Although these boards were small in 1900, they provided a permanent government presence that represented the interests of both sport and commercial fishermen.

A waterman dredging for blue crabs from the Bay—an activity that has declined significantly in recent years. Courtesy the Alliance for the Chesapeake Bay.

Hence, development of the commercial fishery provided the Bay with a twentieth-century legacy. Maryland and Virginia had gone their separate ways in terms of fishery management. And they were to continue to squabble. Some of the most heated and involved disputes were focused on the location of state boundaries in the vicinity of Tangier Island and along the Potomac River. Another hardy perennial dispute was the debate between Maryland and Virginia over the latter's practice of allowing the winter harvest of female crabs.[9] There were also frequent discussions about the effects of the harvests of finfish on the supplies of both commercial and sport species. Maryland, for example, imposed a prohibition on the purse seining of menhaden, arguing that menhaden were the bait fish that attracted large numbers of more desirable predator fish to the upper Bay. Virginia, on the other hand, allowed this efficient method of harvest of menhaden, and a thriving menhaden fishery was established in the lower Bay.[10]

Criticism of Maryland's public oyster fishery continued into the twentieth century. A 1933 conference restated the allegations of mismanagement of Maryland's oyster resources.[11] An official argued that pollution was a "scapegoat" for overfishing. The Baltimore papers, in numerous issues, chided the Maryland General Assembly for failing to capitalize on the productive potential of the Bay. The oyster industry was described as being on an economic par with businesses like "condensed milk and window shades," rather than being recognized as the source of vast wealth that was its heritage and its potential. But the Maryland oyster lobby was politically entrenched. It would continue to be successful in its efforts to rebuff these criticisms; it also stood ready, if need be, to fight new battles in support of the welfare of the oyster industry.

Finally, federal intercession had resulted in formation of the Bay's first bureaucracy. The fishery boards which had been created in Maryland and Virginia had but few powers, but their successors were to play an important role in addressing Bay pollution.

8

Sewage and Shellfish

One of the greatest questions for the future is that of pollution. The pollution of sewage of our ever increasing population and the waste from our rapidly growing industries is affecting the entire fish and oyster industry in and around Hampton Roads. . . .

—Annual Report, Commissioners of Fisheries
of Virginia, 1918–19

As the cities of the Bay region grew, they had to deal with the twin problems of water supply and municipal sanitation. Springs and small streams gradually proved inadequate to supply water for the growing populations, which by the latter part of the nineteenth century began to seek the advantages of indoor plumbing. Now the Bay towns, like cities elsewhere, formed municipal water companies to distribute water through a central system. Richmond and Washington were able to use the large flow of the James and Potomac rivers, while Baltimore reached out into the surrounding country to impound small rural streams. The Norfolk area was forced to rely almost entirely on wells. By the 1870s, all the Bay's cities were using substantially more water per capita, not only for domestic use but for street cleaning, fire protection, and industry.[1] This water, laden with wastes, quickly reached the Bay. By the end of the Civil War, the relatively sluggish tidal exchange rates of the Patapsco basin in Baltimore, the upper Potomac estuary in Washington, and the small tidewater creeks off Hampton Roads caused these waters to be directly offensive to the senses, at least at certain times.[2]

Concurrent with municipal growth was the concern for controlling contagious or epidemic diseases. Prior to the last decade of the nineteenth century, the prevailing theory had been that miasmas, or foul air, transmitted most of the diseases then highly feared, such as cholera, dysentery, and yellow fever. Special attention was, therefore, given to municipal drainage, to the elimination of swamps, and to the design of sewers, so that the latter would not result in the buildup of sewer gas. Contaminated water had, of course, also long been recognized as a source of disease, although the causal agent had not been determined until the germ theory of disease was established just before the turn of the century.[3]

As the result of this interest in water supply, wastewater removal, and public health, the cities around the Bay formed public works bureaucracies starting with municipal water companies, and then going to sewerage commissions.[4] At about the same time, local and state agencies were formed to study public health issues, apply quarantines, and police sanitary practices.

THE BALTIMORE CASE

By 1900 the condition of the Baltimore basin was a longtime topic of agitated debate. In 1893, a sewerage commission was appointed to tackle the problem afresh, since the reports of earlier commissions (1862 and 1883)[5] had not led to action. The commission dutifully set out to study the various alternatives, which it did in great detail, making use of some of the leading authorities of the time. In 1897, it submitted its report,[6] which recommended the building of a sewer that would discharge the untreated wastes of some 350,000 people into the Bay. This approach, which would be unthinkable today, was labeled as the "water carriage-dilution method" of waste disposal, a rather euphemistic term for what another writer vividly called the "construction of a great artificial intestine and anus" for the city.[7]

Had the commission done its careful work just a few years before, its recommendation would likely have been acted on. However, in October 1893, an event took place in Connecticut that directly and dramatically affected the outcome of the Baltimore sewer issue and established what was to become the dominant Bay water-quality issue of the next fifty years. A group of Wesleyan University students became ill with typhoid fever after having eaten oysters.[8] Medical science, armed with the newly accepted germ theory of disease, firmly established that the oysters were the direct cause of the disease. The incident received worldwide medical and popular attention and sent a shock wave through the Bay oyster industry. For the first time, what had been a mere theory, scorned by many—that oysters could cause disease—was now a proven and widely known fact.[9]

No wonder then that the oyster industry, represented both by watermen and packers, voiced forceful objections to the discharge of Baltimore sewage to the Bay.[10] Thus, when the sewerage commission submitted its 1897 report, its entire argument was designed to offset the objections of the oyster interests and to demonstrate that the proposed discharges to the Bay would not cause harm. The commission pointed out that the only incidents of disease came from oysters taken from confined and poorly circulating waters, whereas the proposed discharges would be rapidly diluted and circulated by the immense volume of water in the middle of the Bay. It argued that the cities along the Susquehanna were discharging the wastes of over one million people to the Bay without ill effect. Furthermore, the oyster beds closest to the proposed outfall had been seriously depleted or eliminated by floods of fresh water from the Susquehanna, hence reducing even further any chance of contamination. Indeed, the Bay was productive because of the runoff from the land. The sewerage commission stated: "Your Commission has been advised by an acknowledged authority on all that related to the biology of the oyster that a discharge of the sewage of the city, as here contemplated, would be beneficial rather than injurious to the oysters themselves."[11]

Speaking with the assurance born of careful deliberation and sweet reason, the commission concluded:

> There would appear to be but little reason why the City of
> Baltimore should deny itself the facilities and advantages
> which nature has vouchsafed to it namely the diluting effect of
> the Bay, since the Susquehanna and other towns are doing it
> anyway. It would be ridiculous to attempt to prevent it. . . .
> May not Baltimore, as well, without offense to others, purify
> herself in the broad waters of this great bay without thereby
> disturbing or annoying any existing interest? Your
> Commission thinks she may.[12]

Other, and more potent, interests thought not. Although they apparently did not resort to rhetoric, and, therefore, have left us no written record of their argument, the oyster factions prevailed on the mayor and council to reject the commission's proposal. The local medical community was also involved, although apparently of split opinion, one group arguing that there was no medical risk, the other arguing that pathogens could survive for long periods of time in both saltwater and oysters. And finally, there were the objections of a small but well-organized interest group, the night soil contractors, who made their living collecting the wastes from vault toilets and selling it to area farmers.[13]

These forces combined to persuade the mayor and council to reject the proposal of the commission. In his 1897 message to the city council, the mayor pointed out that the city had a strong economic stake in the oyster industry and, therefore, should not approve a discharge to the Bay.[14] He urged the adoption of "the land filtration technique," an option discussed by the commission but rejected because of its high cost relative to direct discharge to the Bay. This proposal called for pumping the sewage to the Glen Burnie area in Anne Arundel County, a few miles to the south of the city. There the wastes would be used to irrigate and fertilize farmland, a practice widely used in Europe since about the middle of the century. The mayor urged prompt action, exclaiming of the project, "it must be done!"[15] The council agreed to disagree with the commission, and, in 1898, the mayor and council issued a joint resolution rejecting direct Bay discharge.

The commission obligingly went back to work and issued a second report in 1899.[16] When the same members met, they clearly had not changed their opinions as to the soundness of their original recommendation: They spent some time in bolstering the arguments for direct Bay discharge. To their previous list of arguments they added the observation that much waste from Baltimore reached the Bay anyway: through exchange between the waters of the Patapsco and the Bay; through the disposal into the Bay of material dredged from the Baltimore harbor (an observation that presages one of the major issues of the 1960s and beyond); and from runoff from farms that used Baltimore city night soil. Furthermore, argued the commission, if people wanted to worry about the sanitary conditions of oysters, they should worry about oysters harvested from the shallow waters near towns on the tributaries of the Bay. These towns, although not named, surely included Cambridge, Crisfield, and Annapolis, all of which had sizable watermen populations. With a combination of petulance and stiff-necked pride, they concluded their defense of the dilution approach by saying:

However, if our fellow citizens are not inclined to avail of this the dilution method, the natural and most economical method of disposal, and should determine to have something else, leaving the economies of the dilution method to be enjoyed by our neighbors, . . . nothing remains for your Commission . . . than to point out such other methods as seems to it available, though it be much more costly to construct and operate.[17]

With that final blast, they turned to the remaining choices and recommended that the best approach was to treat the sewage and then discharge it to the Bay. Other alternatives combined various treatment strategies and disposal locations, as well as the use of Glen Burnie land disposal for at least part of the sewage from the southern portion of the city.[18]

For political and economical reasons, the city did not act on this set of recommendations. It remained for the great Baltimore Fire of 1904 to clear the way, both physically and politically, for the construction of a comprehensive sewer system. By that time, all other major cities in the country had done so. But by being last, Baltimore vaulted into the lead in sewage treatment, as we will describe. The city government quickly turned to the Maryland General Assembly to get the authorization to form a sewer authority and issue bonds for the construction of a system, an authorization required by the state constitution. The General Assembly gave its approval, but in so doing stipulated that ". . . said commission, to be formed under the Act, shall have no authority to construct and establish any sewage system involving the discharge of sewage, as distinguished from stormwater or ground drainage, into the Chesapeake Bay or any of its tributaries."[19] The oyster interests, in clear control of the matter through their heavy representation of tidewater senators and delegates, thus prevented any possibility of the city making use of the Bay for direct discharge.

Faced with this limitation, the newly constituted commission set about once again to examine the alternatives and come up with a proposal. The first issue was how to interpret the restriction placed on them by the legislature. A strict reading would have suggested that no discharge, treated or untreated, was permitted. Since this was clearly impossible (except through a total land disposal system), the commission, by resolution, adopted the following approach: ". . . that

87

the effluent proposed to be discharged into the Chesapeake Bay or its tributaries in the system to be recommended by the engineers shall be of the highest practicable *degree of purity.*[20] [Emphasis added.] The commission turned, for the fifth time, to the leading sanitary engineers of the Western world. With amazing speed and efficiency, in view of the indecisiveness of the previous forty years, they adopted a plan to build a sewage treatment plant on Back River. The plant was to use sprinkling filters, settling basins, and sand filters, an unprecedented undertaking. Said one of the consultants: "No city in our country of the size of Baltimore has as yet been required to give its sewage treatment with the object of obtaining the highest practicable degree of purity."[21] He added: "We are aware that the works proposed at Baltimore constitute by far the largest undertaking of sewage purification in this country." Indeed, when the plant was put into

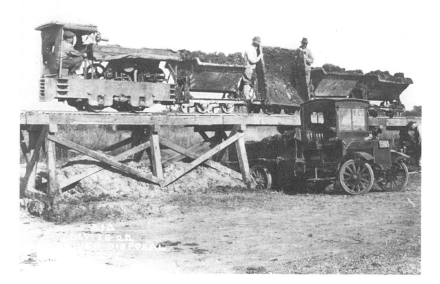

Opening in 1909, Back River Disposal Plant was part of a system which gave Baltimore the best sewerage in the nation at that time. In this photograph, taken by Alfred Waldeck on October 18, 1922, sludge is being dumped into a truck. Courtesy Enoch Pratt Free Library.

operation in 1912, it was regarded as one of the engineering wonders of the modern world.

In adopting this plan, the commission was careful to point out that:

> Owing to the small flow of upland water and the limited range of the tide, there are no strong currents in Back River, and the effluent so discharged would not, under usual conditions, reach its mouth for several weeks. We are of the opinion that there would be no danger from this discharge to the oyster beds of Chesapeake Bay, and no offense along the shores of Back River or elsewhere.[22]

The proposal was also checked "with oyster authorities" and, approved, proceeded apace.

It is perhaps difficult for the modern reader to appreciate the unprecedented nature of the commitment to treat the sewage of a major city discharging to an estuary. Not only was Baltimore the first major city in America to adopt a waste treatment system, but also its reason for doing so was far removed from those commonly used to justify, or require, the treatment of sewage. The strongest impetus for sewage treatment was probably where such treatment would help to protect a city's own water supply. For example, Chicago both discharged its wastes into and drew its drinking water from Lake Michigan. A second incentive was that a downstream neighbor drew its drinking water from the river used by another to dispose of its wastes. A third was that an acute nuisance prevailed, one that for some technical or economic reason could not be solved by piping the wastes away from the city. A fourth was that the disposal of wastes was clearly injurious to fish, recreation, or other resources. None of these reasons applied to Baltimore. Had the wastes been piped raw to the middle of the Bay, no water supplies would have been affected, no nuisance conditions created, and, most probably, no major change to the aquatic life of the Bay induced. Yet, pathogenic bacteria might find their way from a diseased human being to the Bay, thence to an oyster, and finally to another human being. So Baltimore spent an enormous amount of money to treat its wastes, before discharging them into a sluggish tidal tributary of the Bay. As we will discuss, it was not until the late 1930s for Washington and the 1950s for Norfolk that the other two Bay metropolitan areas put sewage treatment plants on line.

The Baltimore sewer story is significant also because it gives a well-documented picture of how city officials or, more particularly, their study commissions and consultants, went about developing their recommendations. The commission reports of 1897, 1899, 1905, and 1906 reflect a remarkably efficient, flexible, and sophisticated approach to the problems. They should be required reading for anyone—professional, citizen, or politician—involved in contemporary sewage treatment controversies, such as the perennial debates involving sewage collection and treatment in the Washington metropolitan area. For they are useful models of how to bring together the best available technical assistance, evaluate the recognized alternatives, reach a reasoned conclusion, and present a written report, all in a relatively short time.

VIRGINIA

While Maryland was going through the resolution of a major water-quality issue in the form of the Baltimore sewage controversy, Virginia was affected in a more insidious way by the aftermath of the Wesleyan University typhoid incident. As the first effect, local boards of health in the Hampton Roads area began to question the safety of, and then closed to harvesting, some of the leased oyster beds in the shallow creeks tributary to the James River and the Bay. Although the number and extent are unknown to us, these closings appear to have been relatively small beds, though highly valuable. Official reaction from the board of fisheries was to sneer at those who were agitating for closure of the beds: "We cannot command language strong enough to denounce the action of some 'pure food' faddists: The scare of 'polluted oysters' has cost the workers in Virginia 'not less than one million dollars a year for 3 or 4 years.' " So said the board in the annual report of 1910.[23] By 1914, it was stating that the waters of Virginia were almost entirely clean, but that because of adverse publicity, the industry was being hurt. In Virginia, as in Maryland, the concern appeared in the early days to be almost entirely about public opinion. The board pointed out that it was difficult for a Bay native to realize that "inlanders" were indeed worried about oyster purity because of the "scare journalism" they were exposed to.[24]

The next year, the commission's tone changed appreciably. Although still labeling it "The Pollution Scare" in its annual report, it was feeling the direct effect of the marketplace, and it suggested action:

> While the polluted area is small . . . so great is its value, and so damaging its existence to the entire industry because of adverse publicity, that we deem it advisable to recommend to the General Assembly the consideration of legislation which will result in the removal of the cause of the pollution. This can be done by the towns installing sewage disposal plants.[25]

This rather indirect statement of inclination, if not intent, at least shifted the problem from the "faddists" to the pollution. No doubt the commissioners were aware of the sizable investment made by Baltimore to treat its wastes. They were at least willing to suggest that it be considered as well in Virginia.

The commission also quoted, with approval, several more emphatic statements about the need to address the source of pollution in tidal waters. Dr. Hugh S. Cumming, a surgeon with the newly formed U.S. Public Health Service, whose extensive surveys on the Bay area are discussed below, wrote to the chairman of the commission: "It is now recognized that no community has the right to dispose of sewage in such a way as to throw unreasonable burdens upon other communities."[26] Although this seems at first inspection to be a simple restatement of a long-established common-law principle, it was cited by the commission as a symptom that times were changing. After giving passing mention to the fact that Maryland had, in 1914, given its board of health authority to require municipalities to install sewage treatment plants, the commission quoted from a paper submitted by a Connecticut official before the National Association of Shellfish Commissioners in 1914: "The contamination of tidal waters . . . is an evil that ought to be stopped. . . . It is a crime against nature, and it is the more indefensible because it is committed, not by savages . . . but by a highly cultured people."[27] Clearly the commissioners did not wish to speak with such rhetorical passion themselves, but they were trying to make a point.

At the same time, the private oyster interests, in the persons of S. J. Watson and Frank W. Darling, sought legal remedies. Watson held a private oyster lease for some eleven acres of bottom in Hampton Creek, within the city limits of Hampton. In 1909, he was informed by county health officers that the waters were "too polluted to permit the sale of oysters therefrom," and in 1914 the state authorities, following recommendations made by a survey of federal health officials, expressly banned the sale of oysters from the creek. The ban affected other planters in the area, and it appears that Watson brought suit on behalf of all of them against the city of Hampton, claiming that its discharge of sewage was unlawful and constituted a trespass against his property, for which he should be compensated. The circuit court ruled in his favor, and awarded him $4,500 in damages.[28]

The case was quickly appealed, and the Supreme Court of Appeals of Virginia reversed on June 8, 1916, in a decision containing a number of important statements and principles. The decision hinged on whether the city, in dumping its wastes in tidewater, could be constrained by the private interests held by Watson. The court held that the city was exercising a public function in accordance with state law, and that this function was superior to any rights granted by the state for private oyster culture. It pointed out that the waters and bottoms of the creek, being navigable and tidal, were "owned and controlled by the state for the use and benefit of all the public." It was, therefore, for the state, "through the legislative branch of government, to say how much pollution it will permit to be emptied into and upon waters, so long as the owners of the land between low-water and high-water mark are not injured. . . ."[29]

Since the legislature had expressly authorized cities to build sewers for the disposal of wastes, and since it had not required sewage treatment or otherwise restricted the disposal of municipal wastes, the city of Hampton was within its rights and was, therefore, not liable for the damages suffered by Watson.

Here is part of the ruling:

> Since the state holds its tidal waters and the beds thereof for
> the benefit of all the public, we are of the opinion that the city
> of Hampton has the right to use the waters of Hampton Creek
> for the purpose of carrying off its refuse and sewage to the sea,
> so long as such use does not constitute a public nuisance. . . .
> The sea is the natural outlet for all the impurities flowing from
> the land, and the public health demands that our large and

rapidly growing seacoast cities should not be obstructed in their use of this outlet, except in the public interest. One great natural office of the sea and of all running waters is to carry off and dissipate, by their perpetual motion and currents, the impurities and offscourings of the land.

The state guards the health of its people for the benefit and protection of the public at large and under present sanitary standards sewerage systems for all thickly settled communities have become an imperative necessity, a public right, which is superior to the leasing by the state of a few acres of oyster land. . . .[30]

In short, the court hammered home the principle that sewage disposal was in the public interest, subject only to standards imposed by the legislature, which governed the use of the tidal waters in the public interest.

Undaunted by the *Watson* decision, Darling brought suit against Newport News for the pollution of Hampton Roads, where he held large and valuable oyster leases.[31] This time the circuit court decided for the city, based on *Hampton v. Watson*, and the Virginia Court of Appeals agreed. The court advanced the strong pragmatic argument that pollution was an inevitable consequence of commerce, industry, and large settlements. Therefore, in providing for private oyster culture, the legislature did so on the assumption that the right of riparians to discharge their wastes to the sea was a superior right. The court said:

The right claimed by the city clearly existed before the enactment of the oyster law cannot be doubted, and the Legislature cannot be presumed to have intended to destroy this ancient and undoubted public right in the absence of a clear and explicit statute indicating such a purpose. We think the more reasonable view of the statute, that established the private oyster program, is that it was not conceived that it would be thought desirable to continue to plant oysters in an area so certain ultimately to be polluted. . . .[32]

In other words, use of waters of the area for oystering is ". . . subject to the ancient right of the riparian owners to drain the harmful refuse of the land into the sea, *which is the sewer provided therefore by nature. . . ."*[33] [Emphasis added.]

Perhaps encouraged by a long and complex dissent written by one of the justices of the Virginia court, Darling proceeded to the U.S. Supreme Court. There, in an opinion handed down on April 28, 1919,[34] Justice Oliver Wendell Holmes made short work of the plaintiff's case and, in sustaining the decision of the state court, added the observation that reflects the prevailing attitudes of the times, as well as the law: "The ocean hitherto has been treated as open to the discharge of sewage from the cities upon its shores. Whatever science may accomplish in the future we are not aware that it yet has discovered any generally accepted way of avoiding the practical necessity of so using the great natural purifying basin."[35]

Ignoring the experience of Baltimore and numerous European cities in the treatment of sewage, Justice Holmes continued:

> But we agree with the court below that when land is let under the water of Hampton Roads, even though let for oyster beds, the lessee must be held to take the risk of pollution of the water. It cannot be supposed that for a dollar an acre, the rent mentioned in the Code, or whatever other sum the plaintiff paid, he acquired a property superior to that risk, or that by the mere making of the lease the State contracted, if it could, against using its legislative power to sanction one of the *very most important public uses* of water already partly polluted, and in the vicinity of half a dozen cities and towns to which that water obviously furnished the natural place of discharge.[36]

These cases, beyond laying the claims of the private oyster lessees to rest, established two things: First, waste disposal was considered to be an important beneficial use of water, a concept that may seem jarring today, but is explicitly stated in much Bay literature and popular writings at least up through the 1950s. Second, the courts made it clear that it was within the power of the Virginia legislature to require a reduction in pollution, though it had not done so.

Litigation out of the way, the situation in Virginia stabilized, so that by 1923 the commissioners of fisheries could say, "As yet Virginia has no serious question of other pollution on its oyster beds, due to the vastness of our waters and no cities or large population within our confines."[37] The "oyster scare" had become a reality for the private planters whose beds were affected, but Virginia at this time had seen fit neither to legislate to restrict waste discharges nor to invest in municipal treatment plants.

FEDERAL GOVERNMENT

Earlier in the twentieth century the federal government had become involved with the Bay fishery. Along with the formation of the Public Health Service in 1912, Congress authorized the conduct of studies on the pollution of navigable waters. This move led to an extensive survey of the Bay by Public Health Service physician Cumming, who described the sanitary condition of the Bay in a 1916 report remarkable for its scope and detail.[38] The report confirmed the commonly held notion that bacterial contamination was largely confined to the waters immediately in the vicinity of populated areas, with the most confined and slowly circulating waters presenting the greatest risk of contamination of shellfish. Cumming, working with health officials in both states, identified a number of specific situations that posed a threat to human health. Many of these dealt with the actual handling of oysters once harvested and with the practice of "drinking" oysters, which involved placing them in fresh or near-fresh water to plump them for the market (a practice which sometimes exposed the oysters to polluted water). The federal presence was, therefore, primarily of a technical assistance and advisory nature, although the Pure Food and Drug Act of 1906 and subsequent acts gave the federal health officials authority to intervene in cases that involved the interstate shipment of contaminated foodstuffs. (Cumming was recognized for his expertise, serving as surgeon general from 1920–36.)

The Public Health Service also had to perform a sanitary survey of the Potomac estuary in 1913–14. The oyster interests on the Potomac—in both Virginia and Maryland—were concerned that the sewage from the growing population of Washington, which in 1913 numbered approximately 320,000, would destroy the purity of Potomac oysters. The survey concluded that there was no immediate risk to the oysters, since the upper limit of the beds was far removed from the Washington sewer outfall, and the net downstream transport of polluted water was slow enough to allow adequate time for dilution and oxidation of the wastes.[39] Thus, the Potomac estuary was spared the problem that was faced in the Baltimore and Hampton Roads areas in the early part of the century.

OYSTER SANITATION PROGRAMS

The stability of the oyster sanitation issue, which is expressed by the 1923 statement of the Virginia Commission of Fisheries quoted above, was shattered by a major outbreak of typhoid in Chicago in November and December 1924, as well as lesser, but significant, outbreaks in Washington and New York. In all, about fifteen hundred cases, resulting in 150 deaths, were reported. Since most of these cases were traced to contaminated oysters, the state of Illinois imposed an immediate ban on the importing of raw oysters. The resulting publicity brought the Bay oyster industry virtually to a standstill.[40]

The Bay states, as well as the other oyster-producing states in the nation, quickly joined with the Public Health Service to establish a program of oyster sanitation that would restore confidence in the purity of oysters. The approach involved the formal adoption of bacter-

As part of the response to public concern about disease associated with Bay oysters, federal authorities inspected incoming takes. In this 1914 photograph, an agent is at work on Long Wharf in the Inner Harbor, Baltimore. Courtesy Enoch Pratt Free Library.

ial standards for oyster-growing waters, as well as standards and practices for the handling and processing of oysters from harvest to the marketplace. Today this system still relies on the federal Public Health Service to certify the adequacy of state programs.

Each state responded quickly to the problem. Maryland immediately moved to close beds in polluted areas, upgrade the sanitation practices of harvesters and packers, and convince the Illinois authorities that the Maryland oyster was indeed safe. To that end, the director of public health for Illinois, Dr. I. D. Rawlings, was brought to Maryland. There he personally conducted an extensive survey of Maryland oyster waters and processing plants. He found that, with a few exceptions, the waters of the state were clean. He suggested that a set of specific practices should be adopted by harvesters, shuckers, and packers to insure that oysters would not be contaminated. By October 1925, the commissioner of conservation was able to report to the governor that the state of Maryland could certify its oysters safe to the state of Illinois, and "that confidence was returning to the industry."[41]

At the same time, beds in suspect or contaminated areas near Cambridge, Crisfield, Salisbury, and Annapolis were closed, and a campaign was launched to eliminate or improve toilets, privies, and pipes that were discharging sewage to tidewater. After a couple of years of effort, the total restricted area was reduced from several thousand acres to 1,288 acres by January 1, 1928, more than half of which were in the Severn River, near Annapolis.

Stimulated at least in part by the oyster sanitation issue, Maryland also launched a sizable campaign to build sewage treatment plants in its tidewater towns. By 1934, it asserted that in the whole state 68 percent of the population was served by sewerage, and that the sewage of 57 percent received some form of treatment, "a record not surpassed by that of any other state in the United States."[42] In that year alone, $4.4 million was spent on sanitary projects, most funded by aid from the U.S. Emergency Relief and Civil Works administrations, two Depression-era sources of assistance.[43]

Virginia also moved quickly in response to the Illinois incident, but its problem and its solutions were markedly different from those of Maryland. Because the center of the Virginia oyster industry was so close to its largest tidewater metropolitan area, the stricter sanitation standards adopted in 1925 resulted in the closing of a much larger area than in Maryland, where, it may be remembered, the largest concentration of people in Baltimore were many miles removed from productive oyster grounds, and the sewage of that city had been undergoing treatment for over a decade. Virginia closed 8,300 acres in 1925, 16,000 more in 1926, and 13,700 more in 1927. By 1933, a total of 38,000 acres were "condemned," of which approximately 12,000 were considered to be productive.

Although Virginia launched a campaign to eliminate contamination in nonsewered areas, it was clear that its primary problem was the untreated sewage discharged by the municipalities, military installations, and other government facilities of the Norfolk–Hampton Roads–Newport News area. Thus, the problem was solvable only through the installation of sewage treatment plants. Such a solution had to gather additional political support and economic feasibility before it was finally undertaken after World War II. Although little concrete action resulted in Virginia, there was a great deal of talk, much of it of considerable historical interest.

As might be expected from previous quotations, the Fisheries Commission confined itself to lamenting the state of things that caused the closing of ". . . the greatest oyster producing territory that ever existed or will exist in the whole world." Although there were, no doubt, unrecorded efforts to address the situation, the commission appeared resigned in 1929, when it said, "It [the Commission] never expects to see Norfolk and Portsmouth provide sewer [sic] disposal plants for the relief of the shellfish industry. These cities may do it in years to come for health conditions and sanitary improvements along their waterfronts."[44] In other words, the cities were acting (or failing to act), and the oyster industry could not muster sufficient strength in the legislature to cause them to do otherwise. The next year, when the bed closings reached 30,000 acres, the commission's apparent passivity matched its concern: "This is a problem so great that it staggers the best thoughts and energies that can be employed. . . ."[45]

The Virginia General Assembly did, however, address the issue when it appointed a commission to study the pollution of tidal waters. This group, known as the Spatley Commission, reported to the General Assembly in 1928. No direct statement of its findings or recommendations was located during this study. Whatever its findings, however, the General Assembly took no action, except that in 1930 it authorized the city of Newport News to issue bonds to construct a sewer and, if the city so chose, a sewage disposal plant, to prevent the pollution of Salter's Creek, the body of water that received most of the wastes from Newport News when Darling sued the city in 1918. The city chose not to build a treatment plant, but proceeded with plans to build a new sewer outfall in Hampton Roads some 2,000 feet from the low-water mark, to replace the main city discharge into Salter's Creek proper. In effect, the city was adopting the dilution strategy: It proposed to move its outfall into a body having a large volume of water and substantial currents to disperse the wastes.[46]

At this point, the executive branch in the person of the attorney general, acting on behalf of the governor, sued the city, requesting the court to require the city to install a modern sewage treatment plant, eliminate its discharge to Salter's Creek, and abandon its plan to discharge untreated sewage into Hampton Roads. That made the suit extraordinary. In effect, the executive branch was suing the General Assembly for failing to exercise its responsibility for protecting the public interest in tidal waters, specifically the rights of public fishing and the right to water clear enough for recreational use. In its brief, the executive branch claimed:

> The General Assembly is the department of the government to which the administration of this property (the tidal waters and bottoms of the state) is committed, but the state as such is the trustee; and no alienation or disposition thereof by the General Assembly is legal which does not recognize and is not an execution of the trust upon which it is held for the people, nor is any use thereof by sufferance of the General Assembly legal which substantially impairs the common rights of fishery or other common rights of the people therein. It is the duty of the General Assembly to make provision for the protection and enforcement of the common rights of the people in the tidal waters. But if it authorizes, permits, or suffers an individual or municipality to use the tidal waters . . . in such a way as to destroy or substantially impair . . . the right of the people . . . it is the duty of the executive department of the state government to invoke the aid of the judicial department to restrain the individual or municipality from so doing.[47]

In a lengthy and fascinating opinion, the Virginia Supreme Court upheld the decision of the Circuit Court of Richmond, which had dismissed the suit (i.e., decided for the city). The court found that the concept of public trust, as argued by the attorney general, did not apply.[48]

The reasoning behind this opinion runs to nearly ten thousand words, and explores some dark and dusty corners of constitutional law, but in sum the court concluded that the right of fishery was something that the General Assembly could restrict or impair, so long as such action was in the public interest; the discharge of untreated wastes into tidal waters, following the arguments of Darling and Watson, was a public beneficial use of long standing; and the General Assembly, by expressly giving Newport News the authority to build a sewer to discharge into Hampton Roads, while neither expressly nor implicitly requiring treatment, "must be construed as authorizing Newport News to discharge its sewage into the Roads untreated." In short, the court reasoned that the benefits of waste disposal and fishery are equal in status. And the restrictions and limitations to be placed on these several uses are questions committed by the U.S. Constitution to the discretion of the legislature, free from control or interference of either the executive or judicial department of the government.[49] Clearly the fisheries interests, both public and private, would have to fight and win their battle in the General Assembly. With closures of oyster beds reaching 60,000 acres by 1934, they must have wished for some of the political clout of their fellow watermen in Maryland, where the oyster interests had forced Baltimore to treat its sewage on the mere suggestion that it might hurt the Maryland industry.

In 1934, the Virginia legislature once again received a report from a pollution study commission, appointed by the governor at the request of the General Assembly "to study . . . the most practical and economic methods of controlling the pollution of [Virginia] waters." The commission submitted a brief and pointed report urging the creation of a "Commission of purification of waters," to be given powers to prohibit pollution, float bonds for the construction of sewage treatment facilities, and approve discharges.[50] The commission was emphatic in its position: "No language used in this report in describing the existing conditions need to be considered as too strong. It is, therefore, urged and *insisted* that the most careful consideration be given to the conditions outlined herein."[51] In outlining the problem, the commission addressed not only the threat to the fisheries industry, but warned that bathing beaches would be forced to close, "property values and tourism destroyed, health threatened, and terrible economic waste created."[52]

The passion of the majority of the commission was more than matched by the unanswered logic of a dissenting member. J. C. Biggins wrote an opinion as long as the majority's and apparently to more effect because the legislature again took no action. His argument could serve as a model for anyone wishing to oppose public investment in waste treatment, because it contained just about every imaginable argument. He first asserted that the public cost would be very high and that there was no guarantee that the situation would be much improved, in that there were numerous sources of pollution for which there were no feasible controls. He cited as an example the situation in Baltimore, where, despite investments of over $20 million, one could still not swim or fish in the harbor. Furthermore, even if substantial improvements could be made, there was a question of whether they would be in the economic interests of the area. Expensive requirements for waste treatment might put the port at a competitive disadvantage relative to more liberal (i.e., permissive) areas. Would it not make more sense, he argued, to recognize a de facto situation and zone the Hampton Roads area for commerce and industry, thereby excluding any claims against these activities by other users (except, presumably, matters pertaining directly to human health or common-law nuisance)?[53]

No doubt the issues were further debated, both in the legislature and without, but we found no direct evidence of that discussion. In 1938, the Virginia legislature created the Hampton Roads Sanitary District, giving this regional body authority to raise money, construct and operate sewage treatment facilities and sewerage systems, and control pollution in the Hampton Roads area.[54] And the "oyster scare" still refused to go away. In their report for 1938–39 the Commissioners of Fisheries lamented that there simply was not enough demand for the oyster crop. Production was not the problem. "There are just not enough people eating seafoods," it said. Here, to be sure, pollution had had its effect. ". . . The thing that is most seriously affecting our consumption in Virginia is the much publicized pollution conditions in Hampton Roads and other Virginia sections."[55] But what was needed was a campaign to convince people to eat oysters as they had in the nineteenth century, and to that end Virginia attempted to develop a marketing exhibit in conjunction with Maryland for the 1939 World's Fair.[56]

This story of sewers and shellfish is a useful paradigm. It determines the importance of the oyster as the key Bay resource. It illustrates also that it is often political influence, not technical evidence, that determines the outcome of an issue. It is the first of many Bay controversies in which the technical questions were fundamentally insoluble. Because it is impossible to demonstrate that a proposed activity will not have an adverse effect, the only available proof is to go ahead with the proposal and observe its effect. Those opposing the change rely primarily on the argument that an adverse effect is possible, and that it is not worth the risk to find out whether indeed such a possibility will be realized. Moreover, the story illustrates that it is perhaps public opinion that is the controlling "fact" in a given issue. The ultimate concern of the oyster interests was not that oysters would be killed or that people would get sick from eating contami-

A patent tonger from Solomons at work at the mouth of Patuxent River. Although this photograph was taken in January 1980, it could have been taken any time within the previous forty years—the hard life of the waterman has not changed. Photograph by Richard I. McLean.

nated oysters. Rather it was that people would associate oysters from the Bay with sewage, make the connection between oysters and sewage-borne disease, and then stop buying Bay oysters. What was controlling was whether *people would think* that oysters from the Bay were tainted.

Finally, this story illustrates the serial nature of environmental problems. The sewer issue began with the adoption of public water systems for the cities. These systems were developed not only in order to supply pure drinking water, but also for the broader purpose of improving municipal sanitation and safety through the provision of water for street cleaning and fire fighting. But problems came with the increased water supply, together with the adoption of the water closet (flush toilet), more frequent bathing, and other amenities of indoor plumbing, all of which can be seen in one context as environmental advances. Contaminated water had to be disposed of. The first approach of the cities was simply to move the wastes downstream and into the middle of the Bay. This approach ran into a competing interest involving another human health issue—the pollution of oyster beds: The remedy became the problem.

9

"Don't Let the Factories In"

Sails are full on the blue bay,
Men are sculling a wooden wind,
Don't take the crabs away.
Don't let the factories in.

—From Gilbert Byron's "Tangier Prayer"
in *These Chesapeake Men*, 1942

The Bay and its tributaries have been used for waste disposal ever since the area was settled. Since the early nineteenth century, sawmills, slaughterhouses, and canneries have used Bay water to flush away their refuse. Statutes were passed in both Bay states prohibiting the dumping of injurious substances into waterways.[1] These statutes were enforced by sheriffs, if at all.

In the twentieth century, as industries grew in size and complexity, and as they tended to become more concentrated, they came to be seen as more of a problem, and thereby drew the attention of government. It was not until after World War II, however, that industrial pollution took on its modern role as a major villain.

The notion of industrial wastes is broad and inclusive. Industries range in the character of their discharges from tomato skins of a canning factory to the complex man-made chemicals of a plastics manufacturing plant. For this discussion of the twentieth century's problems, we will also apply the term to the oil pollution caused by ships using the Bay. Given this range, however, three things stand out about the problems associated with industry: first, oil was seen initially as the major problem for the Bay proper; second, food-processing plants were by far the most numerous and most noticeable sources of industrial problems; and, third, the problems other than oil tended to be associated with free-flowing streams or restricted tidal waters, rather than with the Bay itself, although there are some notable exceptions that will be discussed. The image of a giant industrial complex hovering over the Bay, with huge pipes spewing forth wastes, is new; until recently, industry was a small building employing a few people, and the wastes were shoveled or dumped by hand. In most cases, the wastes were not exotic and lethal; they were mostly familiar, if unsavory, things—like slaughterhouse offal, brewery mash, milk by-products, or chicken scraps.

MARYLAND

Through the nineteenth century, concern mounted over the decline of certain kinds of fish in the Bay and its tributaries. At the turn of the century, the Maryland Commissioners of Fisheries laid the blame squarely on industry, the first such public utterance that we have found:

> We also desire to express our opinion in this report as to the enforcement of local and state laws prohibiting the pollution of streams with chemicals, refuse from canning houses, sawdust, and other stuff injurious to the maintenance of good fishing in our rivers, either by anglers or net fishermen. Complaints have been pouring in upon us for a year or more that the refuse from tomato canning establishments dumped into several rivers were ruining the fishing, and the falling off in the catch dates with the location of the canneries on the river banks.[2]

Since this was a period in which truck farming and canning were at their peak in Maryland, the problems were both relatively new and intense.

The commissioners also commented on the restricted or reduced runs of shad (a topic of concern in both Bay states since about 1830) and identified pollution as one of the probable causes. Public Health Service physician Cumming, in his 1914–16 survey of the Bay, also made note of canning wastes, and quoted an early state health department report that described the foulness of the Cambridge harbor during canning season. "Such a situation is indefensible and should be tolerated no longer by the citizens of the city," it said.[3]

Despite this expression of concern, nothing tangible was done by the state until 1914, when the health department was given authority to control wastes from industries when their discharges constituted a

Unloading a banana steamer, Baltimore. By the turn of the twentieth century, Baltimore had become a busy port, importing from Latin America and the Caribbean. In 1905, when this picture was taken, coffee, sugar, and guano were also imported. The masts in the background rise from Bay boats that brought fruits, vegetables, and oysters for canning, Baltimore then being the canning capital of the world. The debris to the right looks like oyster shells, which often piled up near canneries, but it probably is debris from the fire of 1904 that razed most of the harborside. Courtesy the Library of Congress.

threat to human health or were a nuisance.[4] This somewhat restricted authority was followed in 1917 by a statute that gave the conservation commission authority over discharges that were injurious to fish and other aquatic life.[5] This statute, which appears to have been sometimes vigorously implemented and sometimes virtually ignored, gave the state, in the words of one official, "powers for undertaking the great work of making the waters of the State fertile for the growth of fish and shellfish life."[6]

The conservation and health departments jointly addressed the industrial problems in the 1920s and 1930s and by the late 1930s were joined by another important participant, the Chesapeake Biological

Laboratory at Solomon's Island. The style they developed of dealing with the problems can be characterized as one of cooperation and persuasion, reflected by a statement of the conservation department made in 1922. After outlining the nature and scope of the problem, the report said:

> The Commission does not in any way desire to put any restraint or additional hardship by way of expenditures upon the local capital invested [in industrial plants], but it is necessary that the rivers and streams which empty into the Bay should be cleaned of pollution. . . . Where pollution is found, there will be an endeavor on the part of the Commission to solicit the cooperation of the industry concerned to stop the pollution. By this method we believe pollution can be curbed to a large extent.[7]

The same annual report discussed the state of industrial waste management after the National Association of Fisheries Commissioners had studied the problem for years; they recommended an emphasis on reclamation or recycling. The health department also pursued this approach with considerable vigor, as reflected in its annual reports of the same period. A *Baltimore Sun* feature article written by Pulitzer Prize winner Mark S. Watson in 1930 described the progress being made in Curtis Bay, a largely industrial area south of Baltimore. It described a number of specific cases when the state health department had worked with industries and had suggested ways in which changing processes would both reduce pollution and save money.

> An important part of the work of the State Conservation Department is, of course, the tabulation of statistics about replacement and consumption of the Bay's products. Through Commissioner Swepson Earl's figures we learn that, heavy as have been the inroads on the shad, that fish is far from extinction—for 323,000 were caught in the Bay last year. And with 2,000,000 bushels of oysters produced in Maryland waters in the same period, there would seem to be hope for the oyster industry. And with 209,000 barrels of hard crabs caught in Maryland (plus half as much more from Virginia waters which

the Crisfield packers send to market) it would seem that anyone with a passion for crabs should have been able to gratify it, for crabs run 250 to the barrel, and 200,000 barrels means 50,000,000 crabs. The restaurants from New York to San Francisco have difficulty in getting New England lobsters to satisfy the demand, are using crabs in increasing quantities, so that in fact the Bay crab is gradually replacing the lobster as King of Crustacea.

In such impressive facts as these there is lively demonstration of Maryland's fight against industrialism, for that is what the whole conservation movement amounts to—whether it be protection and restoration of forests, or game, or fish, or anything else. Now, while conservation seeks to aid natural resources, industries often find that it may interfere with their profit making, temporarily at least, and so many of them, through suspicion or ignorance, or plain hostility to any new idea, try to crush it. Hence conservationists use their wits and seek to convince the industries that they will be better off, rather than worse off, through an intelligent conservation program. Thus the natural foes of conservation often are converted into advocates and allies and success comes to the campaigns to save the woodlands from demolition and the water courses from ruinous pollution and their denizens from extinction. Here in Maryland these suasion methods have been signally successful, and the proof of their soundness is at hand.[8]

Curtis Bay continued to be a focal point of Maryland attention in the 1930s. Frequent fish kills and complaints from area residents kept government officials and industries searching for causes and solutions. One of the more intractable problems was a large paint pigments firm, which produced a highly acidic discharge. The health department worked with the company to develop a new process by which the wastes could be reduced. The planned changeover required, as a temporary measure, the discharge of the wastes directly to the Bay. The health department agreed with the opinion of the consulting engineers working for the firm that the discharge would have no adverse effect on oysters and other aquatic life. The conservation department, however, was not persuaded, and "declined permission" for the project.[9]

This situation, which was eventually resolved to everyone's satisfaction by the adoption of the new process, reflected two most significant aspects of the Maryland approach to industrial wastes during this period. First, it is clear from this episode, as well as others like it, that state officials had a powerful influence on industrial activities. While statutory authority was limited and staff small, state officials were directly involved in approving or modifying the industrial discharges. This involvement was a far more activist program than appears to have been appreciated by later students of the Bay. It seems that the primary restrictions on the program were the limited scientific and technical capabilities then available to study specific problems, rather than any fundamental bounds in statutory authority or official concern.

Second, there runs throughout the interplay between the health department and the conservation department two somewhat contrasting approaches to problems. The health department took what for simplicity's sake might be known as an engineering view, while the conservation department took a biological view of the issues at hand. The former tended to be more pragmatic and empirical, stressing doing what was feasible and then seeing what the results were, while the latter tended to be conservative, stressing *potential* damage due to little understood or unknown chemical-biological relations.

Industrial wastes from food-processing plants proved to be generally less amenable to correction by the techniques developed by the Maryland official team. In many cases, there simply were no effective and economical techniques available for dealing with a particular waste. In others, the small size of the operation, its marginal economic status, and, in the case of a cannery, its seasonal nature made it difficult to require installation of an effective waste-disposal system. As a consequence, distilleries, canneries, and milk-processing plants were each identified, at different times, as the major water pollution problem facing the state.

By the late 1930s, however, there was a general note of optimism and progress among Maryland resource officials. With the revival of industrial activity following economic improvement in the mid-1930s, the conservation department found that it had more complaints to investigate. Far from being concerned, the department used this fact as an occasion to report:

> The Department has been very diligent, and in investigating these complaints, [has] notified the owners of these industries that the pollution must cease. The Department is glad to note that in every instance, the officers of the corporations readily cooperated with the Department, and in some instances, have gone to considerable expense to see that no refuse from their factories would, in any way, pollute or destroy the fish life in the waters of the State. The Department feels that they have been successful in protecting the State's waters.[10]

A few years later longtime State Game Warden Lee LeCompte spoke of the great progress that had been made in eliminating industrial pollution, stating that the year 1940 "has been more free from pollution complaints than for at least ten years."[11]

And the health department, in its reports, indicated the activities in the Curtis Bay area which had received the most concentrated attention in the state: "The officials of the industrial plants in the area and the Conservation Commission are to be commended for their splendid cooperation." And of industries: "It has been gratifying to note the recent interest, displayed by manufacturing officials generally, in the matter of stream pollution."[12]

In 1936, the statutory authority of the conservation department was used in court, apparently for the first and only time, resulting in the levying of a $900 fine against the Owings Mills Distillery, Inc., for the release of caustic soda into Jones Falls.[13] The case was hailed as a landmark because it demonstrated the legal backing available should an industry prove recalcitrant. However, the prevailing effort as expressed throughout the period was to resolve industrial pollution problems through cooperation. Speaking generally, but clearly with the Maryland experience in mind, Dr. Abel Wolman summed up this approach as follows:

> If any specific feature has been responsible for progress [in water-quality control] in many of the states, it has been the existence of one or more informed officials who have had sufficient energy and wisdom to carry the program forward by

cooperative activity with industry and municipality. Only rarely in such progressive areas has it been necessary to invoke the law.[14]

By the onset of World War II, the Maryland program of control of industrial effluents and the building of municipal sewage treatment plants was a source of pride. It was the opinion of many that it was a national leader in the field, and it may indeed have been. Certainly, there was a great deal of effective activity by a very small staff, and twenty-five years had seen the establishment of legal basis for action followed by a program that involved the health department, the conservation agencies, and the research establishment in a cooperative routine.

By 1945, the operating question was, "Given the successes of the Maryland program, is it enough?" A number of symptoms appear in the record to suggest that the answer was no. Perhaps the most poignant concern was expressed by R. V. Truitt, the long-term director of CBL (the Chesapeake Biological Laboratory). CBL had been doing pollution-related studies in the Curtis Bay and Baltimore harbor areas since the early 1930s, and by the latter part of the decade was looked to by the conservation agencies as the major source of expert information. Consistently, Dr. Truitt and others at the laboratory had pointed out that the assimilative capacity of the waters of the area was great, and that there was no evidence that industrial wastes were having an adverse effect on economically important resources. In making this statement in 1940, Dr. Truitt stated: "More extensive study is planned on the problem of possible *accumulative effect* of continued discharges of wastes upon *the delicate biological balance of Bay waters* and the relationship of these changes to conservation."[15] [Emphasis added.] Here is sounded, perhaps for the first time, the warning that has been so much a part of the ongoing debate about the quality of the Bay. The CBL scientists were aware of the large quantities of wastes being discharged to the Bay in the Baltimore area, and while they were also aware of the tremendous assimilative capacity of the receiving waters, they also wondered whether other, undetected effects were occurring.

Various groups of citizens were concerned about more palpable issues. Through the first half of the 1940s, a number of incidents and conditions occurred or pertained that raised questions as to the adequacy of the state's program for industrial and municipal pollution control. The residents of Curtis Bay and Back River in particular were not persuaded that progress was being made. Newspaper accounts during that period speak of fish kills, nuisance conditions, and generally unacceptable conditions, and it can be inferred from the degree of official attention to these areas that the level of complaints was relatively high.[16] A 1942 fish kill in Curtis Bay elicited the official explanation that it was part of a broader fish kill that was probably not related to any man-induced condition. In the same period, the health department pointed out that the unpleasant conditions in Back River were not caused by industrial pollution but by "profuse growth of algae due to high organic content of the effluent from the Back River Sewage Treatment Plant." Baltimore's Inner Harbor had once again achieved notoriety as a seriously polluted body of water.[17]

More generally, indeed statewide, the conservation and sportsmen's groups were starting to aggregate themselves for more effective action against industries. In 1935, forty-four conservation organizations and other groups united to form the Maryland Outdoor Life Federation.[18] The federation promptly began to lobby for a change in the organizational structure of the conservation agencies of the state and for the establishment of a board of pollution control. In 1938, a committee on Bay pollution was formed, which pushed for greater recognition of potential and real problems faced by the Bay due to the growth of population and industry. The Izaak Walton League of mostly fishermen kept pressure on for the protection of freshwater streams; it was quick to use fish kills in the Bay or tangible instances of industrial pollution as evidence that a greater level of control was needed. Although the war years tended to reduce the intensity of this pressure, it was there ready to reassert itself thereafter.

VIRGINIA

The Virginia record concerning industrial pollution during the first half of the twentieth century is not as extensive or clear-cut as that of Maryland. Virginia did not have the industrial concentrations similar to those in Curtis Bay or Baltimore harbor. Perhaps partly for this reason, it did not establish a clear legislative framework for action (as did the Maryland statutes of 1914 and 1917). Virginia's concerns during that time were apparently more directed to industrial pollution of inland streams. Nevertheless, a number of specific developments in Virginia during this period concerned industrial pollution, and many of them are important as precursors of the major Bay pollution issues of the 1960s and 1970s. Virginia's experience is, therefore, no less interesting than Maryland's, if somewhat briefer.

Virginia officials had long been concerned over the decline of shad and other anadromous fish in its rivers. These declines were attributed to various factors, including dams, overfishing, and agricultural practices, but there was also recognition that industrial wastes were blocking the passage of fish. Numerous laws were enacted during the nineteenth century to prohibit the discharge of various polluting substances into the streams of the state. For the most part, the language specified free-flowing streams, although it is clear that one of the major concerns was the effects of wastes on anadromous fish from the Bay (and ocean) as well as concern for resident inland fish. These statutes were enforceable primarily by local law enforcement officials; no bureaucracy was established to implement or oversee them. Characteristically, they showed more concern for the deliberate poisoning or killing of fish than for the unintentional effects of industrial wastes, although both categories were covered.[19]

In the first part of the century, the attention that was drawn to the wastes from a pulp mill in Richmond resulted in an extensive study of the situation. This study recommended a combination of civil common-law remedies and new legislation to alleviate what were considered to be unacceptable nuisance conditions.[20] The concerns that prompted this study were primarily those of residents along the banks of the river, rather than anyone's concerns for the fish life in the river. Nonetheless, this study, which resulted in no definite action, represented the first investigation of tidewater industrial pollution located during the course of the research for this book.

With the marked increase in shipping in the lower Chesapeake, the problem of oil from ships came into prominence and for a number of years was seen as the most serious pollution problem in the Bay. The oil came from the practice of ships pumping ballast water from their bilges as they entered the Bay to take on cargo. The problem in the lower Bay was aggravated by the presence of a large number of navy ships. Also, ships heading for Baltimore would begin to pump their bilges as soon as they entered the protected waters of the Bay from the open ocean.[21]

The concern was felt by resort and property owners along the shores of the lower Bay, as the many fine beaches of that area were occasionally fouled by oil. But it was the Commissioners of Fisheries who expressed the strongest opposition to this practice. Although their attention had been directed to the oyster sanitation issue for some years, in 1919 they were to say:

> One of the greatest questions for the future is that of pollution. The pollution by sewage of our ever increasing population and the waste from our rapidly growing industries is affecting the entire fish and oyster industry in and around Hampton Roads, *but of all the destruction caused by pollution that from oil waste is the worst.*[22] [Emphasis added.]

They pointed out that oil, unlike sewage, affects not only mature oysters, smothering them or making them unpalatable, but it also interferes with the plankton that is the food for oysters, and directly kills oyster spat when it is part of the plankton. (Plankton are floating

Deerfoot Cook, chief of the Pamunkey Indians, shows off his catch of shad. Shad and other fish were the major source of income of this Virginia tribe as late as 1960. Courtesy the A. Aubrey Bodine Collection, Peale Museum.

plant and animal organisms that are largely incapable of motion; oyster spat larvae are part of this plankton until they settle and attach to some hard surface on the bottom.)

Thus the commission pinpointed oil as a clear enemy for the organisms of the Bay. This concern was shared by the other East Coast states, including Maryland, and their combined action led to the passage of the Federal Oil Pollution Act of 1924.[23]

How effective this act was is not discernible from the records we examined. Its passage, though, coincided with the "Illinois incident," which quickly vaulted oyster sanitation ahead of oil as a major concern. The fisheries commission, after commenting on oil pollution for a number of years, gave it no mention in its annual reports thereafter. It did receive mention as a significant problem in a 1935 report by the Virginia Planning Board and it was still a major issue in Maryland following World War II. For this period, however, it was an issue that both states conceded was the responsibility of the federal government. It therefore marks the first instance of a direct regulatory role by the federal government in a water-quality issue of the Bay. This modest beginning, of course, gives no clue as to the major growth of federal involvement that took place in the later part of the 1960s and afterward.

In 1930, a massive die-off of oysters in the York River began a controversy, that persists to this date, about the effects of industrial wastes. A large pulp mill at West Point was suspected of being the cause of the die-off.[24] Government officials in Virginia and their representatives in Washington sought the aid of the U.S. Bureau of Fisheries, which established a laboratory at Yorktown "solely for the investigation of industrial pollution in the York River."[25] The study was carried on for most of the next decade. Although oyster productivity declined in the area, and the pulp mill wastes were "definitely implicated," there was no government action taken to abate the agents of the decline. Government at the state and federal levels apparently felt it necessary to establish a direct link between the discharge and the observed declines before requiring remedial action. Such proof was not forthcoming.

What did emerge from the study was the recognition that the state needed a research capability to address the kinds of questions raised by the York pulp mill, as well as to address the broader questions of management of Bay species. After the federal government pulled out of its York River studies, the Commonwealth of Virginia and the College of William and Mary established the Virginia Fisheries Laboratory at Yorktown, with Curtis L. Newcombe as its director. The laboratory took up from the Marine Biological Laboratory the task the latter had adopted in its last years, which was ". . . to find ways and means for improving the tidewater fisheries resources of Virginia. . . ."[26] Thus, while the oyster die-off on the York remained perplexing, the trade wastes of the pulp mill led chronologically, if somewhat tangentially, to the state laboratory, later renamed the Virginia Institute of Marine Science, that was to play a major role in pollution issues after the war.

Virginia also faced an industrial issue in the late 1930s and early 1940s, the resolution of which freed it from difficulties that were later experienced by Maryland. The issue was dredged spoil disposal, a perennial matter of concern on the Bay. The approach channels to Hampton Roads and the port areas required periodic dredging, both to remove accumulated sediment and to increase their size and depth. Prior to the 1930s the practice in the lower Bay had been to dispose of the dredged material overboard in the shallow areas of the Bay, where it would be distributed by the currents and eventually settle out. Watermen and property owners began to object to this activity, claiming that it caused adverse effects either through mechanical smothering or by transferring polluted bottom sediments from the harbor area to the open Bay, where it would contaminate oysters and foul beaches.[27]

In 1940 Congress directed the U.S. Army Corps of Engineers to study the problem and develop a solution. The corps came back in 1944 with a proposal to dike a large area on the south side of the James River just to the west of the Elizabeth River opposite Norfolk. The diked area would receive the dredged spoil from all the public and private navigation projects in the area. At a public hearing on the subject in 1944, some nearby oyster lessees objected that the effluent from the spillways draining the project would pollute their beds. On the basis of this objection the corps required that the state cancel the nearby oyster leases and hold the federal government harmless for any damages done to oyster grounds in the vicinity. The project was approved by Congress in 1946 and built between 1954 and 1957.[28]

By so resolving the issue of dredged spoil disposal, Virginia interests were able to eliminate it as an important issue thereafter, thus avoiding one of the most vexing and time-consuming controversies that have faced state and federal officials in Maryland from the early 1960s to the present.

The experiences of the two states with industrial pollutants shifted the perspective of officials and scientists with respect to the Bay. Pollution had hitherto been seen as an engineering problem. The questions had been how to keep the shipping lanes free or where to place sewage outfalls. But as ever-increasing industrial discharges killed fish, both states established fishery laboratories. Scientists at these laboratories looked at the Bay as a biological system.

As early as 1933 a consensus had developed that the Bay must be considered a single resource unit. In October of that year, an interstate conference on the Bay was held,[29] the first of many conferences to discuss the management of the Bay. The conference proceedings are of interest in that they give a reasonably accurate picture of what was on the minds of the resource managers of the time. Most strikingly, a number of speakers sounded themes very common today: that the Bay is a national treasure, that it must be viewed and managed as a single system, and that some sort of interstate body should be established to deal with the management and protection of the Bay. Although not all speakers addressed these issues, they were stated often enough to suggest that such ideas were by no means rare or radical.

By the early 1940s the perspective of the Bay as an ecological system was well established. Canadian scientist A. G. Huntsman, who was working at Maryland's Chesapeake Biological Laboratory during this period, expressed it well.[30] He suggested that the Bay was an extremely complex and open system, and that long and careful study would be necessary before it could be managed intelligently as a unit. In the meantime, however, it would be necessary to have ". . . investigators prepared to brave the criticisms of the academic theorists. . . . These investigators should not hesitate to draw preliminary conclusions from limited facts. . . ."[31] In other words, although science was going to progress slowly in unraveling the secrets of the Bay, the needs of management for scientific guidance would not wait, and, therefore, there needed to be courageous "directed research."

Huntsman also stressed the openness of the Bay system. He discussed the influence of land drainage from the large watershed area draining to the Chesapeake, and pointed out that much of what went into the Bay started its journey from well outside the tidewater area. He also reminded his readers of the ocean connection, saying: "The Bay is very far from being a discrete productive unit, since there is said to be extensive movement of the fishes out from and back into the Bay."[32] Of man's activities, Huntsman propounded a rather striking view, one related directly to the controversy then under way with regard to the effects of the York pulp mill effluent:

> It is perhaps axiomatic that the changing character of the fisheries of the Bay is determined mainly by changes in the physical chemical conditions of the Bay. But we are prone to think first of man's actions, of the psychology [physiology?] of the fish, and of biological factors, and to overrate the probability of these being responsible for any changes in a fishery.[33]

By 1945 significant advances had been made in efforts to control industrial wastes. Particularly in Maryland, government officials convinced industry to make in-plant changes to reduce discharges. And there had been numerous attempts to address the difficult scientific, technical, and economic problems posed by industrial effluents and their control. But despite this activity, a clear opinion evolved that the situation was growing steadily worse. Growth in population and industrial activity had simply outstripped the watchdog abilities of the governments of the two states. Whether there were actually less or more wastes reaching the Bay in 1945 than in 1900, our research would not allow us to say. Whatever the facts, there was definitely a growing concern among the citizenry of both states, particularly among sportsmen's groups. Pollution of the Bay, and more specifically of many of its tidal and nontidal tributaries, was seen as a major and growing problem.

A Bay Bureaucracy

Rapid growth of population and industry is outstepping the ability of the State Water Control Board . . . to deal with the problem.
—Commissioners of Fisheries of Virginia, 1959

World War II diverted the attention of the body politic away from the Chesapeake Bay. In its aftermath, however, both states created new administrative agencies and charged them with responsibility for controlling pollution on a sustained basis.

When the war ended in 1945, Virginia had in place the Hampton Roads Sanitary Commission, and this commission immediately resumed its work to sewer and to treat municipal and industrial wastes in this prime Virginia trouble spot on the Bay. By 1947 the commission had committed $12 million to the task; and other improvements to municipal systems, actions by industries, and the correction of some of the smaller sources of pollution in restricted waters resulted in the reopening of oyster beds.[1] At the same time, the Virginia Fisheries Laboratory resumed its study of the biology of the Bay; it also turned its attention to exotic pollutants such as DDT.[2]

The most important development of the period occurred far from the Bay in the vicinity of Front Royal, Virginia. A plastics plant had been constructed on a branch of the Shenandoah River, and its discharges, begun in 1940, had completely killed off the aquatic life on a sizable stretch of the river. Fishermen and conservation groups in the area were outraged; they proceeded to seek political support for government action. An influential state delegate from that area, A. Blackman Moore, agreed to serve on a study committee to investigate what needed to be done about water pollution in the state. His committee came forward in 1945 with a recommendation that a state agency be established with responsibility for the management of pollution control throughout the state (excluding the area under the jurisdiction of the Hampton Roads Sanitary District). The bill went before the 1945 session of the General Assembly, and under Moore's protection and influence, passed the House and Senate with only a few dissenting votes. One of the features that allowed for broad acceptance was a grandfather clause that exempted industries and municipalities already in place. Any expansion or modification, however, required regulatory approval by the Water Control Board. The basic approach of the board was to deal with the condition of the stream or receiving body of water, not with the discharge itself. According to the prevailing dilution theory, this approach allowed for the maximum beneficial use of the water for waste disposal as well as other uses. The board, therefore, had to establish that there was a harm to other uses caused by waste disposal before taking direct remedial action.[3]

Both the creation of the Water Control Board and the progress of the Hampton Roads Sanitary District met with enthusiasm from the Marine Resources Commission. In its 1948–49 report, it said, "The problem of pollution is, we think, a vanishing one. . . . We can envision the time in the not too distant future when pollution will no longer be a problem in Virginia, and valuable oyster ground formerly condemned for use will be restored to production."[4]

The next few years seemed to support that optimism. The treatment plants put on line in the Hampton Roads area allowed for the opening of nearly twelve thousand acres of oyster ground, a success that was hailed as a national model.[5] The Water Control Board, using a similar approach in other parts of the state, concentrated on the development of primary treatment plants for municipal systems. Industries were encouraged or pressed to adopt internal process changes to reduce their loads to tidewater and feeder streams.[6]

Through the 1950s, however, the optimism of the Marine Resources Commission eroded. Read in sequence, their biennial comments on the pollution issue sound like a record winding down. In 1953, they said: "The problem of pollution continues to be a serious one. However, the State Water Control Board is making progress in the abatement of pollution for certain streams and the prevention thereof in others."[7] By 1957, the statement was a bit stronger: "The problem of pollution continues to plague the seafood industry despite the fine work of the state Water Control Board and the Hampton Roads Sanitary District. The discharge of industrial waste . . . poses a real problem."[8] By 1959 it was: "Pollution is a continuing and rapidly growing serious problem which requires prompt attention. . . . Rapid growth of population and industry is outstripping the ability of the State Water Control Board and Hampton Roads Sanitary District to deal with the problem."[9] And, by 1961, the commission lamented:

> Contamination of our natural waters by pollutants of various types is one of the most pressing problems facing our Commonwealth today. Pollution is directly affecting our marketing and the consumption of oysters and clams. It is very possible that it may be one of the reasons for our decline in production.[10]

This time there was no mention of the otherwise fine effort of its sister agencies.

A number of factors contributed to this decline in confidence. There was a marked growth in industrial activity in the tidewater area, particularly in the Hopewell area of the James River. Moreover, the research work of the Virginia Fisheries Laboratory identified new concerns. The laboratory investigated pollution sites and fish kills in the early 1950s and reported on the difficulty of doing research in marine environments, when most of the national research upon which standards and pollution control were based had been done on freshwater bodies of water.[11] In its 1956–57 report, it presented a laundry list of things to worry about, while stressing that there was too little knowledge upon which to base sound and enforceable control requirements. It warned of the problems of subtle effects on nursery grounds and spawning areas; of detergents and industrial chemicals that passed through municipal treatment systems unaffected; of toxic insecticides and weed killers that could have major and persistent effects in small concentrations; and of new and poorly understood chemicals, the toxic effects of which had not even been guessed at, let alone studied.[12] More generally, it described the need to study and understand the Bay as a system, in order to be able to assess the effects of man's activities. "There must be more research on pollution," the laboratory urged, ". . . before the Tidewater area undergoes an industrial explosion. . . ."[13]

One problem that drew scientific attention was a massive oyster die-off in the Rappahannock River in 1955, repeating an earlier die-off in 1949. Once again the watermen applied common sense and ar-

A University of Maryland scientist aboard research vessel *R. V. Orion*, off Calvert Cliffs in January 1982, taking a sample from the Bay bottom. He is using a hydraulically operated clamshell dredge to collect sediment, which he will then analyze for toxic metals, radioactivity, and pollutants. Photograph by Richard I. McLean.

gued that it was caused by upstream pollution. The scientific evidence, much of it gathered and evaluated by the Chesapeake Bay Institute, pointed to naturally occurring low oxygen levels brought on by heavy rainfall. In describing the low oxygen conditions found in various parts of the Bay, the Virginia Fisheries Laboratory stated, "There is no evidence whatsoever that pollution contributes significantly to any of these situations."[14] The director of the Chesapeake Bay Institute, in commenting on the results of its studies, expressed sympathy with the watermen in finding it difficult to accept that the problem was a naturally occurring one, since they had seen other periods of heavy rains in which such a die-off had not occurred. The evidence, however, still did not point to any man-induced changes.[15]

In Maryland, support for a pollution control agency had been developing ever since the late 1930s, especially from the conservation and sportsmen's groups. In January of 1945, representatives of the Izaak Walton League and the League of Maryland Sportsmen met with the governor and key legislators to discuss problems pertaining to pollution in free-flowing streams. The health department was brought into these discussions, but the conservation agencies, which had undergone some reorganization in 1939 and 1941, were not.[16] The Board of Natural Resources formed its own study committee and joined the private groups, urging the governor to form a study committee to examine the state program and see what changes, if any, were needed.

By late 1945, a Committee on Water Pollution was formed, and a year later it proposed the formation of a commission similar to that existing in Michigan. The commission would be a coordinating body which drew together the various functions and interests of the health department and the fisheries interests within the board of natural resources.[17]

By the time of the 1947 legislative session, the proposal had changed to creation of a new agency, similar in some respects to the board already in place in Virginia. Sportsmen's groups in particular had argued that a coordinating body would not be sufficient. The need, they felt, was for an agency that would have pollution control as its sole function. Against the objections of municipal and industrial interests, as well as government officials,[18] the bill passed, and the Water Pollution Control Commission was established.[19]

The work of the new entity in its early years is well documented in its annual reports. The commission was pragmatic and modest, stressing the twin principles of cooperation and reasonableness that had guided the work of the health and conservation departments in the 1920s and 1930s. The annual reports and other documents assumed that waters had a natural assimilative capacity, and that it was only after that capacity had been exceeded that interference with other uses occurred. It was the duty of the commission to see that such excesses did not occur or, when they did, to take corrective action. Throughout the process, economic reality had to be kept in mind, and requirements had to be ". . . not only technically attainable but also financially feasible, so as not to impose upon the State's industrial and municipal economy an expense which is out of proportion to the benefits sought."[20] The commission stressed that its work was in the best interests of the economy. It pointed out that it was often to the advantage of business to support pollution abatement programs: "An active water pollution control program . . . is a good business policy for the State as a whole";[21] moreover, "no one reading the technical journals of industry could fail to be impressed by the sincerity of industrial management in its efforts to solve the problem of stream pollution."[22]

During this period the regulators spoke well of the regulations. With a cooperative spirit, they set out with a small staff to attack water-quality problems, which were seen to be wastes from canneries, wash from sand and gravel operations, and oil discharges from ships on the Bay. Work on the first two focused on nontidal streams, although most of the operations were in the coastal plain and, therefore, had at least an indirect effect on the Bay. The oil issue was attacked primarily by enlisting the aid of Bay pilots, who were urged to impress upon ships' captains the importance of observing the Federal Oil Pollution Act of 1924. Although prosecution under the act was difficult and the fines small, a real hardship could be placed on a ship by requiring it to remain in port while an investigation was conducted. This tactic apparently helped reduce the incidence of violations, because oil disappeared from the list of primary pollution concerns in the Bay.[23]

If progress was the official view expressed in the annual reports of the Water Pollution Control Commission, the Health Department, and the Board of Natural Resources, the newspaper coverage of the time noted little improvement. For example, Curtis Bay and Back River continued to be sources of criticism and complaints. In 1948, over fifteen hundred residents of Back River assembled to protest water-quality conditions there, and agitated for the city of Baltimore to do something about the Back River Sewage Treatment Plant. The city was examining a variety of alternatives to upgrade the plant when a circuit court judge ordered the city in 1949 to consider direct discharge to the Bay as an alternative. In so doing, the judge seemed to vindicate the Sewerage Commission of 1899, when it had pressed unsuccessfully for this use of the Bay.[24] Industrial discharges continued to be complained about in Curtis Bay. And in 1955 Anne Arundel County took the unusual step of suing the DuPont Company to enjoin it from building a discharge pipe into the Chesapeake Bay, although the project had the approval of the Water Pollution Control Commission.[25]

In the mid-1950s numerous communities in the upper Bay were still discharging raw sewage into the Bay, prompting the formation of the Upper Chesapeake Watershed Association, a citizen's group whose primary mission was to create pressure for the construction of waste treatment plants. Also in 1955, the first major Bay pollution news article appeared in the *Baltimore Sun*, itemizing the various problem areas, the sources of pollution, and the lack of treatment implemented by either municipalities or industry. This influential article as excerpted below suggested that the Water Pollution Control Commission was something less than aggressive in having taken only four industries to court in the eight years of its existence:

Raw sewage is a disgusting subject. Few people would prefer to talk about it if it could be avoided. But in many parts of Maryland today, raw sewage is something that cannot be avoided. People are being forced to live with it, and their objections are becoming increasingly audible. The facts are these.

Twenty-five Maryland cities and towns have public sewers but no treatment plants. The daily outpourings from the hundreds upon hundreds of homes in these communities are dumped raw into the nearest streams.

Another thirty-two Maryland towns, of sufficient size to have significance from a public health standpoint, have no sewers at all. The raw sewage goes into the ground, to the extent that the ground can absorb it: or it is privately piped or channeled to nearby streams, or it overflows from cesspools and stands in fetid low places.

Still another sixteen Maryland cities and towns have sewers and treatment plants, but they are inadequate to meet the demands made upon them, so that some raw sewage either gets only partial treatment or bypasses the treatment plants altogether and flows directly into streams.

To this list of seventy-three towns must be added the Baltimore metropolitan area, where sewer systems have failed to keep up with population growth, and also those Maryland suburbs of Washington which help to pollute the Potomac. And finally, there are institutions like the Perry Point Veterans Hospital, which empties the waste of 1,500 patients and staff into the Chesapeake Bay.

Raw sewage is not the whole problem, either. Into Maryland streams and Chesapeake Bay go the waste materials of big city and small-town production: acid mine waters, toxic chemicals, offal from meat and poultry packing houses, pulp and seeds from canning companies, the washings from dairies, oil, grease, coal dust, pulp fibers, clay particles, and other foreign matter, to say nothing of the trash and garbage that householders toss into rivers and brooks.

Fifty years ago the disposal of human and industrial wastes was no great problem. The streams used for waste disposal could clear themselves of contamination in short order, using nature's own remedies. Today, however, Maryland is too populated for casual waste disposal.[26]

Also during the same time (the mid-1950s) the municipalities of the Bay area were complaining that the health department was being unreasonable in its demands that they install modern sewage treatment facilities. The towns simply could not afford it, they claimed.[27] These protests that not enough was being done joined those of citizens and conservation groups.

Throughout the period, new water-quality concerns were added to the long-standing issues of municipal and industrial discharges. Wastes from boats became an issue that had received mention in the late 1940s, particularly in the crowded and poorly circulating subsidiaries of the upper western shore, such as the South, Severn, and Magothy rivers in Anne Arundel County, and the Back and Middle rivers in Baltimore County.[28] Growing awareness of pollution as an issue was matched by rapidly growing use of the Bay for recreation. Indeed, recreation users formed a tight causal cycle, as the very boaters and residential shoreline owners who were expressing concern about pollution were also seen as sources of the problem.

Another new issue was the use of the Bay for explosives testing. Although not specifically a water-quality issue, the testing of explosives created a number of highly visible fish kills, roiled water, and created a general nuisance that was complained about by citizens and that caused worry among scientists and managers. The official view was that the military activity was of localized effect and did not constitute a major threat to the Bay. Nonetheless, it was so extensive and so prominent that it came to be seen in some circles as a major menace that significantly interfered with the productivity and enjoyment of the Bay.[29]

Acid mine wastes were also a new concern of the 1950s. Coal mining in the upper reaches of the Potomac and the Susquehanna had created hundreds of miles of highly acidic and biologically dead streams. Acidic waters were for the most part buffered by the time they reached the Bay. Popular suspicion, though, held that the evil-looking "yellow boy" in mountain streams would eventually work its way to the Bay.

In 1954, the U.S. Bureau of Mines developed a plan that would have brought mine wastes directly to the Bay. In what was probably the biggest of the "big pipe" proposals for waste disposal in or to the Bay, the bureau proposed that a 180-inch pipe be built from the anthracite coal region of northeastern Pennsylvania to the Bay, in order to drain the underground coal mines that were rapidly filling with water.[30] The water could not have been pumped into nearby streams because of its high acidity, but the bureau argued that the Bay, with its enormous quantity of water, would quickly neutralize the acids in the mine drainage, and suffer no harm.

Maryland officials responded with outrage. After assuring watermen and sportsmen's groups that the state of Maryland would not tolerate such an invasion from a neighboring state, the governor himself held meetings with the congressional delegation and with officials in the Department of the Interior to seek to get the project withdrawn. The alacrity of the response was perhaps unnecessary, as it appeared that the proposal had serious technical and economic flaws. Pennsylvania officials were not themselves enthusiastic about

Culm piles, or coal refuse heaps, showing erosion at Shenandoah, Pennsylvania, 1936. Courtesy National Archives.

the project, so that by the end of the year it had died. Nonetheless, it is a project of considerable interest because it reflects another example of the attractiveness of the Bay as a sink for wastes, and it represents one of the first clear examples of an interstate water-quality issue.

A second "big pipe" proposal marked the era, this one with a source closer to the Bay. Consulting engineers working for the District of Columbia proposed that the municipal wastes from that city, which contained a substantial loading from the Maryland suburbs, be piped across southern Maryland and discharged into the main stem of the Bay. This would remove pressure from the highly restricted and slowly circulating Potomac River estuary and provide a source of irrigation water to southern Maryland farmers, then experiencing a rather severe drought. In addition, suggested the engineers, the wastes would provide additional enrichment to the Bay, which would be beneficial to the total productivity of the Bay and not injure any of the Bay's resources.[31]

Hence the decade of the 1950s was marked by two major developments. Both states created water pollution control agencies and sought to cooperate with industry and municipalities in developing financially "reasonable" pollution-control strategies. These efforts first met with enthusiasm but, as the decade progressed, first conservation groups and then the agencies themselves seemed to lose confidence in their ability to control pollution.

11

Save the Bay: The 1960s and Early 1970s

You and I differ principally in philosophy, and this should not be misconstrued. You are inclined to encourage full use of the Bay unless available knowledge proves that human uses will be impaired. I am more conservative and prefer not to risk damage to the Bay until reasonably good estimates can be made of all effects.
—L. Eugene Cronin in a letter to Donald W. Pritchard, March 6, 1969

Clustered around 1960, a number of events presaged fundamental changes in the kinds of issues that would concern Bay governments. Virtually all of the previous responses had been prompted by episodes when municipal and industrial wastes at various discrete locations attracted the attention of scientists, administrators, legislators, and the courts. After 1960 the venues became larger as the concept of water quality became all-inclusive; governments more and more asked the question: What is the condition of the Chesapeake Bay? The Bay became generally recognized as an ecological system, subject to a wide variety of man-made influences transcending the boundaries of states, the jurisdictions of agencies, and the capacities of science to define and predict.

Some issues of the 1960s had been around for some time; others were new. But all were marked by problems of increased size, complexity, and persistence. Flow regulation or modification of the major rivers (Susquehanna, Potomac, and James) as a tool in water-quality management came under consideration. The effects of waste-heat discharges, particularly from power plants, began to be seen as a major threat to the Bay. Spoil disposal in the Maryland portion of the Bay developed into a full-blown controversy. Overenrichment of the Bay and heads of the major subestuaries became a matter of concern. The effects of navigation projects on salinity, first on the James and then on the old Chesapeake and Delaware Canal, became a point of debate. And large-scale engineering projects raised the question of cumulative effects of all of man's activity. Rapid development in the Washington suburbs and its effect on sewage disposal and soil erosion introduced land use practices as a source of concern. And the benefits that tidal wetlands provide to the Chesapeake Bay became recognized, leading to enactment by Maryland and Virginia of regulatory programs to protect them.

Tying all of these issues together were a series of journalistic accounts of the Bay and its myriad problems. These articles, both news and feature, began in the early 1960s and reached a peak of volume and urgency in the early 1970s. To the detailing of these issues we now turn.

In the early 1960s, the programs of the two state water pollution control agencies operated substantially as before. In Virginia, there was consensus that there had been significant progress on the municipal waste front and a reduction in oil pollution, and that there was a need to work more directly with industries to reduce wastes through the construction of waste treatment facilities. Previously, industries had been urged to adopt internal process changes that would reduce their discharges.[1] In Maryland, programs continued to stress the need for cooperation with industry in setting realistic goals and providing economically feasible solutions. Toward that end, the Maryland commission began a series of annual conferences to discuss waste-control technologies.[2]

Nonetheless, there were signs that the rapid population growth throughout the region was rendering existing programs inadequate. The Virginia Commission of Fisheries, true to form, continued to express concern for the effects of pollution, while praising or at least encouraging the Water Control Board and the Hampton Roads Sanitary Commission. "Contamination of our natural waters by pollutants of various types is one of the most pressing problems facing our Commonwealth today," it said in 1961.[3] And through the decade, while varying its primary concern from industrial to municipal pollution, it was consistent in the opinion that the problem was grave. In 1971 it stated, repeating its 1961 refrain, that "contamination of our natural waters by primary and secondary pollution is one of the major problems confronting the seafood industry. . . . our paramount concern at the present is the problem of municipal sewage."[4] Treatment plants

A man crabbing near the mouth of the Chesapeake Bay, Lynnhaven Inlet, Virginia, July 1975. Photograph by Richard I. McLean.

generally were not able to handle storm loads, and, as a consequence, oyster beds were frequently closed for some time following rains. And in 1972, the commission's position was this:

> We know that the Water Control Board is working diligently to prevent contamination of our waters, but this is a gigantic task. . . . The Commission prays to the General Assembly to hear the plea of an endangered industry and to act with courage and decisiveness in providing the impetus for swift corrective action in this area.[5]

In Maryland, the fisheries industry was optimistic. The areas closed to shellfish were considerably smaller than in Virginia, and the MSX disease that had greatly reduced oyster stocks in Virginia had little effect in Maryland. The 1960s also marked a time of rapid increase in Maryland's oyster harvest, due to an aggressive state-sponsored reseeding program and several years of above-average natural reproduction. Thus, Maryland was able to recapture first place in national oyster production, a spot it had relinquished to Virginia in 1930.[6]

Maryland was also successful in upgrading its municipal waste treatment plants, so that by 1969 it claimed that approximately 80 percent of its Bay area sewage received secondary treatment. And on the industrial front, there was a consistent sense of progress in the reports of the Water Pollution Control Commission (renamed and reorganized as the Department of Water Resources in 1964), where the number of orders for compliance, investigations, court actions, and other measures of activity continued to increase.[7] These official reports are less than convincing, however, since they give only a numerical report on activity, and do not reflect the volumes of discharged wastes or the actual conditions of the receiving waters.

A significant change in the federal role in water-quality control came about with the passage of the federal Water Pollution Control Act of 1964. This act required that the states establish water-quality standards for all interstate waters within their borders, in order to qualify for federal assistance in waste treatment financing and to avoid direct federal enforcement intervention. Both Maryland and Virginia developed programs of compliance and, in the process, underwent rapid growth in their water control staffs and budgets. The direct impact of this administrative and regulatory change on the Bay is difficult to discern, but it is noteworthy that the shift to water-quality standards did not produce a record by which net progress could be measured (despite the emphasis on the quality of the receiving water rather than on control of discharges). The only available benchmark for achievement of conventional water-quality objectives (i.e., control of discharges from industries and municipal sewer systems) appears to be in the summary comments of the numerous journalistic discussions of Bay water quality during this period, which will be discussed below.

SALINITY MODIFICATION

In the early 1960s, two Corps of Engineers' proposals for major public works projects on the James River introduced salinity modification as a "water quality" issue and prompted scientific study and political controversy. The first was a river basin program for the entire river, issued in 1962.[8] The multipurpose program, consisting primarily of a system of dams, would have allowed for the partial control of freshwater flows into the estuary, thus affecting salinity. The second project proposed deepening the shipping channel to Richmond from 25 to 35 feet.[9] Because salinity distribution in the Bay and its tributaries is affected by the size and depth of the channels, this project also suggested that a change in salinity would occur.

Salinity had long been a matter of concern to Virginia oystermen. At the low end of the range of the oyster, an unusual rise in freshwater inflow will depress salinity and result in stress or death to oysters. This change was often experienced on the Potomac and to a lesser extent on the other major Virginia rivers. Generally of more concern, however, were the higher salinities, because numerous pests of oysters were more prevalent in the higher salinities. Other things being equal, it was and is desirable to keep maximum salinities low, especially in the vital seed-producing portion of the James. The invasion of MSX in the late 1950s, which had hit the Virginia oyster industry so hard, underscored this relation. MSX was generally limited to higher-salinity waters.[10]

In considering these two corps projects, then, Virginia officials and watermen attempted to assess the impact of the proposals on the health of the oyster industry. This concern, of course, was nothing new. What was novel was the fact that they were now dealing with a fundamental parameter of the natural system, salinity, rather than a relatively small quantity of a constituent added by man. The concern had enlarged from pollution to basic ecology. Obviously significant changes in the salinity of the James could have substantial effects on the entire biological system supported by the estuary. The question was, what changes were desirable, and what were undesirable?

The evaluation of the proposed river basin plan led to a generally favorable response from the tidewater interests. The result of the program would have been to even out the flow of the river over the course of the year. That evening-out would prevent summer salinities from going too high in the lower estuary, thus holding the high-salinity predators at bay. It also would reduce the heavy slugs of floodwaters, which tended to limit oyster production in the upper part of the estuary. Two related water-quality benefits were expected. Increased summer flows would provide more dilution waters for the wastes entering the river at Richmond. This increase would tend to lessen the degree of pollution downstream in the freshwater portion of the estuary. Second, the reduction in floodwaters would also reduce the oxygen-depleting effects of storms, which had been held accountable for heavy oyster mortalities in the Rappahannock River during the previous decade.

The proposed shipping channel was quite another matter. Largely through the careful work of the Chesapeake Bay Institute, it was understood that the James River, as well as the Bay and the other major tributaries, had a two-layered circulation pattern: a net upstream flow of dense high-salinity water in the lower part of the water column, and a net seaward flow of fresher, less-dense water above. Deepening and widening the channel, it was reasoned, might increase the total volume of seawater movement upstream and result in higher salinities. It would also physically aid the upstream movement of oyster predators. The fisheries interests, faced with a further threat to their depleted industry, vigorously opposed the project.[11]

After a couple of years of jockeying between the fisheries and port interests, a positive approach to resolving the conflict was adopted in 1964. The Virginia General Assembly appropriated $400,000 for the construction of a hydraulic model of the lower James estuary. This device to test the effects of the construction of the navigation

Hydraulic model of the lower James River estuary, built in 1964 and disman-tled in the early 1980s. Pictured here is the Chickahominy River reach. Cour-tesy College of William and Mary, Virginia Institute of Marine Science, School of Marine Science.

channel on salinity distributions was quickly constructed at the Corps of Engineers' waterways experiment station in Vicksburg, Missis-sippi. Using the corps' experience in calibrating and testing such a model, together with inputs from the Bay scientific community, it was determined that the channel would have virtually no effect on the prevailing salinity regimen of the estuary. On the basis of that finding, the governor was authorized by the General Assembly to give consent to the project.[12]

What had been a major standoff between competing interests was resolved by using the best technology available to make an informed judgment as to the effects of the project. The report stands as a landmark of rationality and objectivity in Bay resource management decision making. Regrettably, from the standpoint of completeness, the navigation project was never begun. During the time that the project was in doubt, an oil pipeline was built between Hampton and Richmond. Since oil was one of the commodities that had given the project its economic viability, the provision of an alternative transportation mode changed the picture. Upon reevaluation the corps found the project to be only marginally in the black (with a benefit/cost ratio of barely over one), and so the project was withdrawn. It should be noted that there were other environmental problems associated with the project, which would have become more prominent had it been pursued. Chief among them was the problem of spoil disposal, which had long vexed Maryland officials. Since that problem came to a head in Maryland at about the same time that the James River controversy was under way, we will now turn to it.

DISPOSAL OF DREDGED SPOIL

Navigation channels and facilities in and around Baltimore harbor have required frequent dredging since the early part of the nineteenth century. The Patapsco River, unlike Hampton Roads, is not a natural deep-water port. Also, the Patapsco receives erosion from a 400-square-mile watershed, most of it hilly and highly erodible, and thus contributes much more silt to the harbor than do the small and flowing rivers of the Hampton Roads area.

Indeed, the shallow waters around the Bay have been called not only "its chief source of productivity" but also "the chief impediment to man's use of the estuary."[13] Almost every harbor on the Bay must be regularly redredged to keep it open. The large commercial channels must be maintained at greater depths—30 to 50 feet. By the 1970s, the viability of Baltimore as a major port was at stake. Governor Harry Hughes sounded what was to become the call to arms for greater dredging in 1979. Before groups of business leaders, politicians, labor leaders, and others, he said it was time to "dredge or die."[14]

Dredging won wide support. Just where to dump the mud was the problem. Disposing of the resulting millions of square yards of dredged material in a way that was environmentally sound yet economically realistic had become, and remains, one of the Bay's critical management issues.[15] Each year as much as 5 million cubic yards of sand and silt are dredged from the Bay and harbor channels. As this is written, 90 million cubic yards are expected to come during the next 20 years from the delta of shipping channels at the Port of Baltimore alone. Current disposal sites, built in the face of vehement opposition, are filling up, necessitating a search for new sites, with new controversies likely to arise over each prospect.

In the nineteenth century, most of the materials dredged from the channels was simply transported to a nearby area and dumped overboard. From the beginning, this practice bothered oystermen, who claimed that it smothered their beds. On the other hand, port authorities complained that oystermen damaged the channels by dredging for oysters along their edges. Such, apparently, was the nineteenth-century standoff on the issue.[16]

With the expansion of the port and approach channels in the 1950s, the Corps of Engineers began to dispose of spoil in the Kent Island Deep, an area of deep water that lay to the east of the main channel near the Bay Bridge. This practice was strongly objected to by Kent Island residents and watermen. Gradually it also became a matter of increasing concern to fisheries biologists and managers.

In 1959, after years of receiving complaints on the issue, Governor J. Millard Tawes appointed a committee to study the matter. Although the primary concern was the biological effects of spoil disposal, arguments arose from Kent Island property owners that the filling of the Kent Island Deep was changing the currents along the shoreline and greatly increasing the rate of shore erosion. They also claimed that the water along the shoreline was made more turbid by the dumping. Although both of these claims were considered far-fetched by scientists, they added to the pressure on politicians to act.[17]

In 1961 the governor's committee report recommended formation of a permanent commission to review all disposal questions and other issues involving the use of state submerged lands. It also recommended that the state adopt a policy of land disposal of spoil where possible: "The deep waters of the Bay remain an important ecological environment which should remain free from encroachment whenever possible."[18] The governor promptly formed a Submerged Lands Commission, made up of various state natural resources and economic development interests, and chaired by the state comptroller.

In the mid-1960s, the spoil disposal issues raised by the expansion of the Chesapeake and Delaware Canal took the spotlight away from the Kent Island Deep issue. The canal, which was being widened from 250 to 450 feet, and deepened from 27 to 35 feet, produced an enormous quantity of dredged spoil. Much of that spoil taken from the canal proper was disposed of on land, but the dredging in the approach channels was pumped to nearby shallow areas and to the Pooles Island Deep. The concern of the Maryland fisheries officials was that this spoil would damage the spawning and nursery areas for a number of important finfish, principally the striped bass. It might also put further stress on the salinity-limited oyster beds in the upper Bay, whose oysters had been on the decline since 1881.

Significant attention was given to this issue by both the Submerged Lands Commission and the Board of Natural Resources. For a time it was proposed that all dredged spoil be placed on marshlands on Aberdeen Proving Ground, but the competing state interests reached a compromise that would allow for some disposal on land and some in Pooles Island Deep. The Corps of Engineers, however, would not accept the additional costs involved in such a practice. So eventually the state resources agencies had to back down under pressure from port development interests, and the overboard disposal continued for both the Chesapeake and Delaware Canal and the Baltimore harbor projects.[19]

In the meantime, a number of studies were conducted that tended to diminish concern that the dredging and spoil disposal were having pronounced adverse effects. The primary result of these studies, aside from advancing knowledge of the biology of the upper Bay, was to provide guidelines for the timing of dredging and spoil disposal, so as not to interfere with spring spawning of striped bass. These guidelines were adopted and the spoil issue appeared for a time to be a problem solved by debate and compromise, and by the use of applied scientific research.

In 1968, however, an additional element entered the spoil disposal picture. Research had been going on for about a decade on the uptake of heavy metals by shellfish. For a shorter time, detailed studies had been made of the water quality and sediments of Baltimore

139

harbor. In 1968, the Submerged Lands Commission was officially informed of the problems posed by the high concentrations of heavy metals found there.[20] To dispose of these sediments in the open Bay would raise the possibility that shellfish would concentrate these metals to the point that they would pose a threat to human health. At that time, there was substantial uncertainty as to the actual risk involved. Little was known about actual physical, chemical, and biological routes by which heavy metals might find their way into shellfish. Furthermore, there was little agreement as to the levels of metals that should be tolerated by humans.

Despite these uncertainties, the need for action was politically clear. Here once again was a threat to the oyster industry, even if not in the direct form of a public relations problem that might affect the marketing of oysters, whether contaminated or not. The state moved quickly to develop a solution, one that still ignites political fury thirty years later. The plan was hatched innocuously enough.

Following a study, the Submerged Lands Commission proposed the construction of a contained spoil disposal area to receive contaminated spoil from Baltimore harbor. A number of possible sites were considered; the one found most economical and feasible involved the construction of a gigantic man-made island in the vicinity of Hart and Miller islands just to the northeast of the harbor entrance. (Back in the 1890s the Baltimore Sewerage Commission proposed to build the Baltimore sewer outfall over these islands.) The original 1968 plan called for the island to be approximately 2 miles long, ½ mile wide, and 25 feet above mean water, with a total capacity of 125 million cubic yards.

The plan received the approval of the state resource agencies, the Port Authority, and the Corps of Engineers, all of whom were represented on the Submerged Lands Commission. Here are key words in the proposal, formally adopted by the governor and submitted to the General Assembly in February 1969:

> Confined disposal of certain types of dredging spoil is essential to the preservation of the environmental integrity of Chesapeake Bay. Provision of the contained disposal area will assure the continued orderly development of water-oriented industry of the State without jeopardizing water quality for at least the next twenty years.[21]

The General Assembly quickly approved a bond issue for $13 million to construct the disposal area, and the detailed engineering studies for the project were begun.[22]

With a solution to the contaminated spoil disposal problem in the offing, Maryland resource agencies argued for the suspension of harbor dredging until the Hart-Miller project was completed. Commenting on a Corps of Engineers project in the fall of 1970, the deputy secretary of the Department of Natural Resources, however, recommended disposal, "since the confined disposal area for the Upper Bay should be in operation by the time this dredging project is to be initiated [in the winter of 1971–72]." In 1971 the director of the Fish and Wildlife Administration recommended disapproval of a Maryland Port Authority project in Dundalk, citing the risk that overboard disposal in the upper Bay would pose to a "developing commercial product, and freshwater clam *Rangia cuneata*." Federal fish and wildlife and water-quality officials also became involved in the process, since all dredging projects needed a certification that they would not contravene existing water-quality standards or federal guidelines. What resulted was a series of negotiated compromises that allowed limited dredging while the details of the Hart-Miller proposal were developed, and while the biological questions relating to contaminated spoil were further studied.[23]

The more significant controversy was not between resource and port development officials, or between state and federal levels of authority. The Hart-Miller project, which had been so carefully worked out between competing interests at the state level, and which was seen by the state resource officials as a major environmental victory for protecting the quality of the Bay, was temporarily stalled by determined local opposition. Residents of the two peninsulas of land to the west of Hart and Miller islands found the prospect of a large new land mass objectionable. The reasons for the opposition ranged from objections to destruction of the small natural islands to the claims that the man-made island would change the circulation patterns of Back and Middle rivers and do irreparable damage to the ecology and recreational values of the entire area. Residents objected to the change to the landscape, to the possibilities that the disposal area would smell and would pollute groundwater, and that birds using it would sicken and die from exposure to the contaminated wastes. They claimed initially that the island would eventually be

141

used as an industrial park, a use which would further degrade the area. The state then gave assurances that the island would eventually become a state park.[24]

At the state wetlands hearing in the spring of 1971, several hundred residents turned out to express their opposition to the project, often jeering the numerous state officials who testified on behalf of the project; but the state gave quick approval. On February 23, 1972, the State of Maryland, through its Department of General Services, filed an application with the Corps of Engineers for a permit, under section 9 of the Rivers and Harbors Act and section 404 of the federal Water Pollution Control Act, to construct a dike and dredged spoil disposal site at Hart and Miller islands. As a result, the opponents turned their attention to the federal approval process. Joined by boaters and fishermen who used the existing islands and nearby waters for recreation, they launched a campaign of opposition that stalled the proposal.

A powerful theme running throughout the controversy was one of class conflict. The elite of business, politics, and the news media backed the Hart-Miller project as essential to the commercial viability of the port. The southeastern Baltimore County communities, where the project would send clouds of mosquitoes and raise the prospect of contamination, were inhabited by a working class fearful of the plan. Their congressman, U.S. Representative Clarence Long, led the fight, giving full vent to his constituents' resentment over their status as the state's dumping ground. He explained the politics of it this way:

> In any other part of the United States where there is a beautiful coastline, the wealthy would be living there. But for some reason the working class settled on the eastern coast of our county and never moved. That's the only reason Hart and Miller were considered for this project. People assumed they could dump on the working class with impunity.
>
> It just seems absurd to haul this spoil laden with mercury, cadmium and lead out to a clean part of the Bay and dump it there in a dike that could break and create a dreary wasteland. In the valley [the wealthy areas of western Baltimore County] they don't understand why these people object. But if they proposed a dump in the valley, they'd come out with a battalion of lawyers that could defeat the Russians.[25]

With the divisions of class and politics arrayed thus, the outcome was easy to predict, though Long and his constituents fought a stout rearguard action. Soon after the Corps of Engineers issued a permit for the project, opponents sued in federal court seeking an injunction. But the suit ultimately failed,[26] and the Hart and Miller islands disposal facility finally was constructed between 1981 and 1983 at a cost of approximately $60 million. Dikes rising 28 feet above the Bay waters now enclose a 1,100-acre site that has a capacity for as much as 70 million cubic feet of dredged spoil.

The Hart-Miller controversy is of special interest in this narrative because it represents perhaps the most significant issue in the history of the Bay (at least up to 1972), when viewed from the standpoint of its economic impact. Yet, its origins, in concerns for the marketability of Bay oysters, were largely a matter of conjecture. In 1965 or 1970 no evidence existed of an actual health risk. Certainly no modern equivalent of the 1900 Wesleyan University typhoid incident prompted the attempts to block the Hart-Miller project. A body of scientific opinion suggests, however, that it would be safer to use deep-water overboard disposal of contaminated spoil, rather than a contained area, since the material would be returned to the environment in which it had been stable for some time, in which there would be little biological activity, and in which it would not be exposed to oxygen and the leaching effect of rainfall, as it would be in a contained area.

Having hitched the fate of dredged spoil to contained areas in the hard-fought Hart-Miller controversy, however, Maryland authorities were unresponsive to this new evidence. Building on 1975 legislation limiting dumping of dredged spoil and sewage sludge into the Bay,[27] the Maryland General Assembly extended the prohibition in 1981 to the dumping of spoil taken from the Baltimore County tributaries of the Bay within 5 miles of the county's Hart-Miller-Pleasure islands chain. Since 1991, the deep trough south of the Chesapeake Bay Bridge has also been off-limits.[28]

Meanwhile, the contained areas were filling up. The Hart-Miller facility continued to receive contaminated and uncontaminated spoil from maintenance dredging of the Baltimore harbor channel, primarily from the showcase Inner Harbor, and was expected to be topped off by 1998. Likewise, uncontaminated spoil being dumped at a designated open-water site near Pooles Island was projected to reach the brim by 1999.

Proposed solutions threatened to revive the Hart-Miller controversy and spark similar confrontations in other communities. Deep-
draft ships were coming perilously close to scraping the 50-foot bottom of the Baltimore harbor channel. Digging deeper would require
new disposal sites. One proposal to raise the dikes at the northern
end of the Hart-Miller facility, to extend its useful life up to thirteen
years, has met with opposition from residents and politicians of
nearby southeastern Baltimore County. Residents there claim the
drying cracks of Hart-Miller's north cell provide a breeding ground
for mosquitoes that drift across the water to plague them in their back-
yards. Some golfers have even said the infestation keeps them away
from the course at Rocky Point. And few take comfort in government
assurances that the facility's containment dike is sound and that the
likelihood of a storm breaking it, and releasing tens of millions of cubic
yards of contaminants into the upper Bay, is highly remote. "How can
they say that with guarantees?" asked their local legislator, Delegate
John S. Arnick. "They said the *Titanic* couldn't sink either."[29]

A more comprehensive plan for dumping uncontaminated
dredged spoil at Poplar Island, about 20 miles south of Hart and
Miller islands, was touted as reclamation of lost environment. The
former island, now an erosion-ravaged archipelago, was once the site
of the Jefferson Islands Club where President Franklin D. Roosevelt
ate oysters and terrapin and where President Harry S. Truman later
played poker.[30] The Corps of Engineers proposed to restore these
remnants to their previous mass, circa 1847, of more than 1,000 acres,
landscaping it as a wildlife habitat. The plan envisioned a high
ground planted with trees at one end and a low marsh teeming with
fish, ducks, and birds at the other. Sea grasses growing in a protected
bay on the restored island's eastern shore would offer habitat and
food for aquatic life.

Skepticism and cost concerns, however, stalled this project, at
least for a time. A *Washington Post* outdoors columnist cautioned
against hope for lush habitat at Poplar Island anytime soon. Because
the island would receive dredged spoil for at least twenty years, "it'll
be years after that before vegetation is complete." At best, the columnist wrote, the project may produce an "innovative, moderately inof-
fensive dredged-spoil dump. And don't look too close."[31] The Maryland General Assembly, too, had its doubts. Staring at cost estimates
of $50 million to $200 million, the 1996 legislature stated in its budget

bill that Maryland should consider more modest options instead. Nevertheless, Governor Parris Glendening budgeted $35 million toward the Poplar Island project in 1996, in the hope of drawing federal support for it. That summer, Congress responded and began appropriating money for the island restoration.[32]

Holding fast to the state's preference for contained sites since the Hart-Miller controversy, Governor Glendening continued the state's opposition to deep-water disposal. Industry and labor leaders lobbied heavily in favor of the idea. "Without Deep Trough or something similar to it, we don't have a long-term program for dredging," said Tay Yoshitani, the Maryland Port Authority's executive director. "If the steamship industry believes Baltimore has failed to address the issue, business here could be seriously jeopardized, particularly with ships getting larger and larger." In the statistics of progressively deeper ship drafts there was an echo of the "dredge or die" shibboleth. Sixty years ago, the average steamship draft was 26 feet. Currently ships pass the port's 50-foot deep channel with about 5 feet or less to spare. Larger ships under construction are expected to go deeper.[33]

The governor, however, heeded the warnings of environmentalists and watermen that dumping in the Deep Trough would interfere with Bay cleanup efforts and impair commercial fishing. "That's the only smooth bottom we have to use the kind of gear we're forced to use in Maryland to get striped bass," said Larry Simns, president of the Maryland Watermen's Association. "It's a natural place that blue crab and rockfish like to go."[34]

With the governor and other forces arrayed against it, the Deep Trough option never came up in the 1996 regular session of the Maryland General Assembly.

THERMAL DISCHARGES

If spoil disposal in the upper Bay has been the issue involving the longest conflict between competing Bay uses, the use of the Bay for power plant cooling water was perhaps the issue that developed the most intense controversy. Starting in 1961 with a coal-fired power plant at Chalk Point on the Patuxent River, then another conventional plant at Morgantown on the Potomac, and a nuclear plant on the James River in Virginia, the issue culminated in 1971 with the

Calvert Cliffs nuclear power plant on the main stem of the Bay in Maryland.

Maryland officials had first become concerned about the question of the biological effects of cooling-water discharges while reviewing the proposed power plant on the free-flowing portion of the Potomac at Dickerson. There was an emerging body of literature which recognized that elevated temperatures could alter river biota. When the Chalk Point plant proposal was received in early 1961, a new set of questions was raised because of the estuarine character of the river. The Water Pollution Control Commission therefore established a research program to be conducted by the Natural Resources Institute, a research and teaching arm of the University of Maryland. The major component of NRI became the Chesapeake Biological Laboratory at Solomons.

The concerns of the Maryland officials were threefold. First was the effect of the exposure to high heat on organisms taken into the power plant condensers. Second was the effect of the chlorination that was necessary to keep the condensers free of algae and other fouling organisms. Third, and most important, was the effect of the heated-water discharge on the ecology of the estuary in that area. Since the plant caused a substantial temperature rise in the water withdrawn for cooling purposes, and used a relatively large portion of the available water for cooling, it was at least possible that the plant would have a major, and possibly negative, effect on the local aquatic life.

The research progressed over a number of years, first involving collection of background information, then making comparative studies once the plant went into operation. As the results were completed, a generally unfavorable picture began to emerge. The Natural Resources Institute's scientists implicated the plant in a number of adverse conditions they observed, most notably a large crab die-off in 1964, a die-off of finfish the same year, and an increased incidence of "green" oysters in the vicinity of the plant, which they attributed to copper coming from the plant's condenser tubes.[35]

The conclusions of the Natural Resources Institute were generally challenged by officials in the Department of Chesapeake Bay Affairs, who felt the issue of green oysters was exaggerated and threatened to raise a new worry in the minds of the consuming public. Scientists hired by the power company put forward reports that argued that the plant had little untoward effects. The Department of Water Resources generally accepted the research of the power company, and even later suggested that the objectivity of the Natural Resources Institute group and its work was subject to question. Scientists at the Chesapeake Bay Institute, who also were conducting research on power plant discharges, generally felt that the effects of the Chalk Point plant were minor and, for the most part, temporary. Virginia Institute of Marine Science scientists, on the other hand, were generally disposed to take the same view as the Natural Resources Institute scientists, that the power plants posed substantial risks to the Bay environment.

This variance of professional and scientific opinion might have been of little long-term account, had it been restricted to Chalk Point. In fact, however, the results of the Natural Resources Institute's work were being released and discussed in the press during the same period in which Maryland also was evaluating a power plant proposal at Morgantown, a nuclear power plant was under construction on the James, and the Baltimore Gas and Electric Company announced its intention to build a nuclear power plant on the Bay at Calvert Cliffs. In a short time, thermal pollution became a major new issue, and the basic disagreements over Chalk Point expanded at Morgantown and Calvert Cliffs.

The exact sequence of events relating to the three power plants is now virtually impossible to sort out, particularly since much of the early consideration of the thermal issue went unrecorded. However, the principal initiator of the public controversy was the announcement by Baltimore Gas and Electric in May 1967 that it intended to build a massive nuclear power plant on the Chesapeake Bay. Many different interests opposed this proposal. The first concern was for the cliffs themselves, since the proposed plant was at the site of extraordinary fossil deposits from the Miocene period. Second were the concerns of property owners in the vicinity of the plant, particularly those whose properties would be traversed by the transmission lines carrying the electricity to the Baltimore area. Third were concerns about the risks of radiation, nuclear waste disposal, or accidents. And, finally, there were those who concentrated on the possible ill effects of the waste heat discharged into the Bay.

Although all these concerns received considerable publicity, and perhaps the land use issue generated the most active opposition, it was the thermal question that received the most attention from officials and the press. By early 1969, enough public concern had been raised to result in six bills being introduced in the Maryland General Assembly regarding power plants on the Chesapeake, including one that would impose a five-year moratorium until a study could be conducted on the effects of waste heat discharges.[36] None of the bills passed, but opposition to Calvert Cliffs continued to grow.

In the meantime, the Department of Water Resources had granted a permit for the operation of the Morgantown facility, an issue that had been under study for over two years. Primarily because of the controversy over Calvert Cliffs, and fed by the reports of NRI from its Chalk Point studies, the Morgantown action also drew heavy fire, some of it from the Virginia side of the river.

Prompted by these objections, the director of the Chesapeake Bay Institute, Dr. Donald Pritchard, issued a statement in which he discussed the likely effects of the Morgantown facility on the Potomac. His reasons for doing so were explicit:

> The considerable newspaper coverage of statements concerning possible excessive damage to the Chesapeake Bay and its tributary estuaries from thermal effects of electric power plants now under construction or planned, and the expression of concern by the members of the State legislature, has led to the general public impression that the Chesapeake Bay and its tributary estuaries are in imminent danger of catastrophic damage to commercial fisheries, sport fisheries, recreation, and other uses. It is my firm opinion that such is not the case. . . . In my opinion, the public as well as their elected representatives have been misled as to the scope of the problem, and the degree to which existing knowledge can be utilized in appraising the problem. . . . A number of scientists and pseudoscientists in this state have been quoted, and misquoted, as predicting severe damage from thermal pollution. Those of us who have contrary views have so far remained publicly silent. I find it necessary to break this silence.[37]

He then went on to describe the Morgantown project in the light of the Natural Resources Institute findings on Chalk Point, and in the process leveled a direct criticism of the findings of its research and the conclusions drawn from them. In a cover letter to a state senator, he summarized his opinion on the NRI work:

> I am of the firm opinion that nowhere in the report published by the Natural Resources Institute . . . is there a demonstration that man's use of and harvest from the waters of the Patuxent estuary adjacent to the Chalk Point Power Plant has been adversely effected [sic] by excess temperatures resulting from operations of that plant.

He ended his report with the unequivocal statement:

> I must conclude that the operation of the Morgantown Power plant, under conditions prescribed by the Department of Water Resources, will have no measurably effected [sic].[38]

Dr. Pritchard's statement was quickly followed by one from the director of the Natural Resources Institute, Dr. L. Eugene Cronin. In it, Dr. Cronin made a point-by-point response to the issues raised by the Chalk Point study. Then he summarized his major concerns, with which "the biologists of the Virginia Institute of Marine Science and members of the Potomac River Fisheries Commission concur[red]":

> a. Allow no elevation above 90 degrees F.; b. Prohibit the use of "tempering water" [a practice of drawing additional river water into the discharge canal so that the final temperature would not exceed the prescribed maximum]; and, c. *The biological studies by company-hired consultants are poorly designed, contain serious errors and are completely inadequate as a baseline of present conditions or for predictions of future effects.*[39] [Emphasis supplied in original.]

That Dr. Cronin saw fit to emphasize the last point makes it clear that this was a major source of concern and disagreement.

He then went on to characterize his difference with Dr. Pritchard:

> You and I differ principally in philosophy, and this should not be misconstrued. You are inclined to encourage full use of the

Bay unless available knowledge proves that human uses will be impaired. I am more conservative and prefer not to risk damage to the Bay until reasonably good estimates can be made of all effects.[40]

He concluded by quoting Dr. Pritchard's final statement to the effect that there would be no adverse effects from Morgantown (quoted above), and commented on it thus:

With the best will in the world, I must say that the fragmentary information available is totally insufficient as a basis for such an absolute statement, which does not sound like your usual precise and well-qualified thinking. . . . Perhaps it is easier for physical oceanographers to make such flat predictions than for biologists. I would never make them, and feel that it is of the utmost importance that we all avoid sweeping conclusions.[41]

These two statements, issued by the two leading Maryland scientific policy advisors, provide a striking outline of the type of problem faced by Bay officials since about 1960. Dr. Pritchard laid particular stress on the public relations aspects of the controversy. Dr. Cronin stated a basic difference in point of view ("philosophy," in his words) that would lead to very different government policies and actions, depending on which was adopted. And behind the debate was the question of the reliability and objectivity of the scientific findings developed by various laboratories and individual researchers.

Faced with these unresolved questions, Governor Marvin Mandel, at the request of the General Assembly, appointed a commission to study the matter. Under the chairmanship of Dr. William W. Eaton, the committee examined the record both as to the thermal and nuclear issues. In late December 1969, it issued its report, saying in effect that the Calvert Cliffs plant, if built under the standards imposed by the Department of Water Resources and the Department of Health and Mental Hygiene, would not pose a threat to the Bay.[42]

The commission was clearly impressed by the economic argument that the Bay represented a valuable resource as a cooling medium. It pointed out that the entire Maryland portion of the Bay would only be elevated 1.5°F in temperature by thirty power plants the size of Calvert Cliffs (assuming even distribution of the waste heat), and it concluded therefrom: "Research to find what fraction of the potential several hundreds of millions of dollars cost of artificial cooling systems can be avoided through Bay use hardly needs economic justification."[43] In March 1970, Dr. Cronin and the principal investigator at Chalk Point, Dr. Joseph A. Mihursky, issued a press release in which they itemized a number of unanswered questions regarding the effects of the plant and suggested that a strong research program be adopted to address those questions. If the plant is approved, they said, and is proven to have adverse effects, penalties should be imposed and immediate corrective action taken.[44]

The document also included additional statements by Drs. Cronin and Pritchard. The former stated his basic precept: "It is of the utmost importance that the effects of large environmental changes be estimated with reasonable accuracy prior to decision and commitment on those changes."[45]

The latter commented:

It is an unfortunate fact that much of the research on the effects of thermal discharges into natural bodies have [sic] been biased, even though possibly unconsciously, towards either showing that no damage occurs, or is intended to prove the conservationist's opinion that catastrophic damage will result from such discharges. While it will be difficult to assure that all research personnel are free from some bias in this matter, particularly in view of recent events here in Maryland, steps could be taken to assure an unbiased review and research guidance procedure.[46]

In May 1970, the Department of Water Resources issued to the Baltimore Gas and Electric Company its final permit, which allowed the appropriation of roughly 3.5 billion gallons of water per day for passage through the plant as cooling water. This permit disposed of the principal water-quality hurdle at the state level, and was followed in 1971 by a Certificate of Public Convenience issued by the Maryland Public Service Commission, a write-off by the U.S. Environmental Protection Agency to the Atomic Energy Commission, a Maryland Wetlands Permit, and eventual approval by the Atomic Energy Commission itself. (It should not be thought that approvals were straightforward and without difficulty. The numerous legal and administrative issues involved in this project would take a book-length manuscript to detail, including the landmark decision by the U.S. Circuit Court of Appeals in July 1971 that found the Atomic Energy Commission inadequate in its fulfillment of the requirements of the National Environmental Policy Act. The details surrounding the various permits issued by Maryland alone occupy several file drawers.) The Calvert Cliffs plant began operation in 1974, and in 1976 produced more electricity than any other nuclear power plant in the free world.[47] In 1980, the company claimed that $400 million had been saved as compared to generating the same amount of electricity using oil as fuel.[48]

Countering the alarm over the prospect of harmful thermal discharges, scientific studies in the late 1970s and early 1980s generally concluded that power plants, including the Calvert Cliffs reactor,

The Baltimore Gas and Electric Company's Calvert Cliffs nuclear power plant. This aerial photograph shows the entire complex except for the discharge station, which is off to the upper right. The circular containment buildings house the reactors. In the foreground is the intake embayment area that draws cooling water from the Bay. Courtesy the Baltimore Gas and Electric Company.

have had little adverse effect on marine life in the Bay.[49] As of 1996, no new scientific studies had emerged to challenge these findings.

While this major conflict was erupting in Maryland, Virginia went about the business of approving the Surry nuclear power plant on the James River at Hog Island with little fanfare and no visible controversy. In the words of the director of the Water Control Board, it "went through as slick as a whistle."[50] The absence of controversy itself presents a problem in analysis, because there is much less in the written record from which to judge why things were so different south of the Potomac. The Virginia Commission of Fisheries did comment in a general way that thermal pollution was an additional threat to the Bay, but there was no overt attention to the Surry plant. One fact that no doubt contributed to the smoother approval process was that the hydraulic model for the James was available to test the physical effects of waste heat discharge. Working under contract with the Virginia Electric Power Company, scientists were apparently able to satisfy Virginia government officials that these thermal effects would not have adverse biological consequences. That the whole process in Virginia occurred several years before the Morgantown and Calvert Cliffs issues in Maryland probably made for a smoother approval. No in-state research had alleged negative effects from other power plants: the Chalk Point work in Maryland was just emerging as the Surry plant was going through its final approvals in 1967. Also, the state permit process was significantly less cumbersome in Virginia, so that decisions were more focused both in place and time. Perhaps most significantly, there was no concentrated neighborhood or conservation group opposition, as in Maryland, and little comment on the project by the press. By the time the Baywide press began to treat wastes as a major issue, the Surry plant had already been approved and was well along to completion. Thus, the issue that perhaps steamed the tempers of quantities of people in Maryland, if not the waters of the Bay, caused nary a mild fever in Virginia.

INTERSTATE ISSUES

The Susquehanna River exerts enormous influence on the salinity and purity of the upper Bay. Without the Susquehanna's daily infusion of 25 million gallons of fresh water, constituting 90 percent of all fresh water entering the upper Chesapeake, the Bay's extraordinary mix of plants and animals would be reduced to that of a mere saltwater inlet.[51] Unfortunately, the Susquehanna also brings pollution in the form of soil nutrients from the agriculture that flourishes outward from its banks. Clearly, any attempt to clean up the Bay would require the cooperation of the states that control the vast Susquehanna watershed, from which both life-sustaining fresh water, and pollution that threatens it, so abundantly flow.[52]

Recognizing the need during the 1960s for collaboration in the future health of this vital artery, New York, Pennsylvania, and Maryland—the states through which the Susquehanna flows—worked to form the Susquehanna River Basin Commission. Its task was to devise a compact mechanism for addressing the management of interstate issues.

Because very little of the Susquehanna is in Maryland, officials there were chiefly interested in addressing the relation of the river to the Chesapeake Bay. In the mid-1960s, various officials of Maryland resource agencies participated in discussions considering regulation of flows in the Susquehanna. They took the same stand the Virginians had taken with regard to the James: general support for a regime that would reduce flooding and increase minimum flows, in the belief that such an approach would increase the range of the oyster while reducing the range of its principal enemies. In addition, reduction in floods would reduce the bacterial contamination that every year required the closing of the upper Bay to shellfish harvesting in the spring months.

Much of the interest in the Susquehanna in the early 1970s arose from the severe drought experienced by the eastern United States from about 1961 to 1966. The Susquehanna, as the largest eastern river discharging to the Atlantic (albeit indirectly, through the Chesapeake Bay), was looked upon as a potential water supply not only within the basin, but also in the adjacent Delaware and Potomac basins. During the same period, greatly increased use of river water for irrigation was being developed, and power plants were being planned that would require substantial amounts of water for cooling.

After years of preliminary discussion, the Susquehanna River Basin Commission was created in 1970 amid great hope that finally a body was imbued with the plenary power to guard the watershed and intervene in the problems of its use and pollution. The commission, however, turned out to be little more than a faithful generator of reports and studies that made little impact. Its aims and actions were so modest that for years its headquarters north of downtown Harrisburg, Pennsylvania, lacked a sign out front. Meanwhile the disparate government agencies the commission was supposed to bring together in a mighty collaboration continued to make decisions piecemeal and in isolation. Susan Q. Stranahan, a *Philadelphia Inquirer* writer covering the Bay, captured some of the general dismay with the commission's "toothless" performance: "The one governmental body that could have monitored the cumulative impact of all those decisions—and challenged those not in the best interest of the entire watershed—remained mute."[53]

For example, the commission could have insisted early on that farms in Lancaster County, Pennsylvania, control the erosion and nutrient runoff that contribute so much to the pollution swept into the Bay by the Susquehanna, but it didn't. "If the Commission had done its job, there would have been no need for the Chesapeake Bay program here in Pennsylvania," said Maurice K. "Doc" Goddard, Pennsylvania's longtime secretary of forests and waters. "Instead ... they just laid back and said nothing about anything."[54]

By the end of the decade, issues concerning the Susquehanna were being eclipsed by concern over the diversion of fresh water from the upper Chesapeake to the Delaware Bay by use of the Chesapeake and Delaware Canal. The issue began to take form in January 1968, when Dr. Pritchard, while working on a Delaware estuary problem, discovered a Corps of Engineers study done in 1934 that estimated that there was a net outflow from the Chesapeake Bay to the Delaware Bay of slightly less than 1,000 cubic feet per second. Since the canal's width and depth were being enlarged from 250 by 27 feet to 450 by 35 feet, Dr. Pritchard reasoned that the net outflow would substantially increase. He saw this as a factor perhaps more significant to the physical and biological properties of the Bay than any other diversions of the Susquehanna's flow then under consideration. He made this possibility known to Maryland officials and other scientists, and began his own calculations on the probable effect of the canal's enlargement.[55]

The issue was quickly picked up by government staff workers, citizens' conservation groups, watermen, and the press. By early 1969, the Chesapeake and Delaware Canal had joined the list of major Bay problems, along with Calvert Cliffs, wetlands destruction, and spoil disposal, in addition to the usual municipal and industrial discharges. Expression of concern ranged from the carefully couched statements of Dr. Cronin, who said that the effects could not be predicted, but might well be significant, to the plaint of a waterman who suggested that the canal be made big enough so that instead of having a dead bay (Delaware) connected to a half-dead one (Chesapeake), we could have just one big dead bay.

In 1970, at the urging of members of the Maryland Congressional delegation, the Committee on Public Works of the House of Representatives held hearings on the matter. Ostensibly the committee was to consider whether the canal project, which had been authorized in 1954 and was already largely completed at a cost of $65 million, should be continued.[56]

Testimony was varied. The committee was considerably disturbed by the recommendation of a senior Department of the Interior official that the project be stopped until exhaustive oceanographic and biological studies could be performed. Members expressed their dismay that a federal agency would come forward at such a late date and make a case against a project that had been available to public and official scrutiny for so long. The Interior Department official responded that new information justified taking a fresh look at a large-scale project of this kind, and that the value of the Chesapeake Bay was so great that a cautious approach was called for.[57]

The key testimony came from Drs. Pritchard and Cronin, who issued statements arguing that the probable effects of completion of the project would be small, but that studies should be conducted that would provide adequate physical and biological understanding of the Bay so that the long-term effects of the enlargement could be assessed. They suggested that such studies would indicate whether there was a need for corrective measures (such as locks), if adverse biological effects could be attributed to the canal.[58]

Following a recess of several weeks, the hearing reconvened with testimony from the Corps of Engineers and Drs. Pritchard and Cronin, outlining a research program to address the issues raised previously. The committee responded favorably to the testimony, with the general sense that the project could proceed with concurrent research. The possibility that research results might indicate the need for locks was again raised, but was not addressed in any detail.[59]

Three months later, a $798,600 research contract was awarded to the University of Maryland to conduct studies of the upper Bay.[60] That work was conducted over the next several years, as was mathematical modeling by the Corps of Engineers to refine the estimates of the physical response of the system to the enlargement of the canal.

The report, issued jointly in 1973 by the marine institutes of the University of Maryland, the Johns Hopkins University, and the University of Delaware, concluded that widening the canal would have no significant effect on salinity or marine life. Experience has so far borne out this finding. No major changes in the environment of the upper Bay that would be attributable to the enlargement of the canal have been observed since the project was completed in 1975.[61]

By the mid-1970s, therefore, the Chesapeake and Delaware Canal had faded from discussions of Bay water quality. Thus, in the space of just a few years, an issue was defined on the basis of initially obscure and largely technical data, was elevated through official and journalistic attention to the status of a crisis, and passed quickly and quietly to oblivion, to join such issues as salinity modification and explosives testing as items removed from the Bay political agenda.

COMPREHENSIVE POINT OF VIEW

The 1960s marked the beginning of systematic attempts to address Bay governance from a comprehensive point of view. The 1933 Chesapeake Bay Conference had been the first such attempt, but it left little evidence except the document that records the conference. Beginning in 1960, however, both Maryland and Virginia and several federal agencies conducted studies or planning efforts that purported to be comprehensive, and in 1971 there was a brief attempt to address the interstate aspects of Bay governance. Bay water-quality matters, although not the only elements of these efforts, were certainly the dominant ones, and were linked to most of the others.

The first activity was a study of water resource management in Maryland by the Board of Natural Resources. Although not specific to the Bay, it addressed a number of Bay issues, and made as its summary finding a recommendation that water management be structured at the state level so that the relations between various functional areas could be addressed.[62] This work was followed by a major study of water management in Maryland that led in 1964 to a wholesale reorganization of the major natural resource agencies in the state. In particular with regard to the Bay, the Department of Tidewater Fisheries was reconstituted as the Department of Chesapeake Bay Affairs, with the responsibility for developing a comprehensive plan for the management of the Bay, together with its more traditional duties of managing the sport and commercial fisheries of tidewater Maryland.[63]

Also during the early part of the decade, the Public Health Service was developing a basinwide perspective on the Susquehanna-Chesapeake system. Starting in 1963, it established a formal Chesapeake Bay–Susquehanna River Basin Project, which was scheduled to run for six years and involve a staff of up to seventy-five persons.[64] The stated purpose of the project was to perform basic water-quality studies and develop a basinwide framework within which water-quality and other management decisions could be made. The study staff saw its mission as bringing into sharper focus the relation of the free-flowing river to the Bay.[65] Of particular interest was the question of enrichment of the Bay, which some of the researchers considered to be the most important water-quality problem for the Chesapeake.

After an ambitious start, the project became the victim of changing priorities within the federal Water Pollution Control Administration, which came into being in 1965 and took over the water-quality functions of the Public Health Service. The project completed a number of specific studies, but never published anything approaching a comprehensive discussion of the basin.

Meanwhile, the numerous political issues on the Bay that involved questions of the effects of engineering changes raised the idea that a physical model for the entire Bay would be a useful research tool. Various members of the Maryland Congressional delegation particularly were impressed by the range of Bay problems and the need to address them in an integrated way. In 1965, the Congress authorized the Corps of Engineers "to make a complete investigation and study of water utilization and control of Chesapeake Bay Basin . . . including . . . navigation, fisheries, flood control, control of noxious waste, water pollution, water-quality control, beach erosion, and recreation."

To aid in this study, the corps was authorized to build an 8-acre hydraulic model of the Bay in Matapeake, Maryland. Sheltered within a 14-acre building, huge pumps churned at one end of the model, simulating the tides, as nozzles gushed from the remaining sides in emulation of the Bay's tributary rivers. The Susquehanna and Potomac nozzles emitted a steady stream while smaller ones dribbled as creeks. Known as "the Monster of Matapeake," the concrete mold prototype was 1,000 feet long and 21 inches deep at its deepest, holding about 450,000 gallons of water. As the contraption worked, scientists could be seen stepping over its rivers or crossing the Bay part over long catwalks. The idea was to replicate the dynamics of freshwater inflows, tides, and temperature; the model could run a year of Bay water movements in three and a half days. At a cost of $15 million, the project was supposed to save money and time over the old way of monitoring the Bay's behavior by measurements taken from boats over the course of weeks, months, even years.[66]

The hydraulic model saw only limited use as a research tool, however, between 1973 and 1983. Its last big experiment involved determining where flotsam and bodies might turn up after a commercial airliner crashed in the Potomac in January 1982. After that, the corps abandoned the model because advances in computer simulation had made it obsolete. Asked if the model was a white elephant, Dr. Eugene Cronin acknowledged, "It certainly has some light gray streaks in it."[67]

The corps study took approximately ten years, resulting in the publication of the multivolume Existing Conditions Report in 1973 and Future Conditions Report in 1977.[68] At the time, these reports represented by far the heftiest, if not the most illuminating, discussion of the complex relations between various Bay functional areas. For all its detailed documentation of the impacts of future growth patterns on the Bay, however, the reports were largely ignored by government agencies regulating water quality, probably because the results were only descriptive of problems and entirely lacking in proposals for corrective action.[69]

Several other federal studies focused on the Bay in the latter part of the 1960s. Both the National Estuarine Pollution Study of the federal Water Pollution Control Administration and the National Estuary Study of the Fish and Wildlife Service had special sections on the Chesapeake Bay that discussed the complex interfunctional and interstate aspects of the various water-quality problems they identified. The National Council on Marine Resources and Engineering

Development had a task group that made a case study of the Bay in order to formulate policy for the coastal zone.[70] And NASA, eager to demonstrate its earth-resources capabilities, established several projects and held a symposium on the applications of space technology to the management of the Bay.[71]

Of all the high-level studies of the Bay, the one that produced the quickest, most far-reaching results was an informal, haphazard one that took place a few years later. Charles "Mac" Mathias, then Maryland's junior U.S. senator, gathered all the data he needed in the mid-1970s while sailing from cove to cove along the Bay's great length to observe environmental conditions. He was appalled. Returning to Washington, he managed to insert in the EPA appropriation bill a directive for a comprehensive EPA study of the Bay. The legislation established the federal Chesapeake Bay Program, directing the EPA to (1) assess the principal factors harming the Bay environment; (2) analyze all environmental data being collected on the Bay; (3) establish the continuing capability for collecting, storing, analyzing, and disseminating such data; (4) start a sampling program; (5) determine which government agencies have responsibility to manage environmental quality of the Bay; and (6) find ways to improve Bay management.[72]

Fulfilling the mandate of the Mathias legislation of 1975, the EPA undertook studies for seven years at a cost of $27 million. Five EPA reports, issued in 1982 and 1983, concluded that the Bay environment was declining in several ways, particularly owing to the runoff of fertilizers from farmland. The reports, discussed at length in the next chapter, composed what has since become known as the Chesapeake Bay Study.

The states were also active during the latter part of the 1960s. Maryland held a Governor's Conference on the Chesapeake Bay in September 1968 for the purpose of laying "a foundation for the orderly development of Chesapeake Bay." The conference posed basic policy questions, most of which remain unanswered. Tom Horton, then the environmental writer for *The Baltimore Sun*, in 1987 looked back wistfully on the conference as an early indicator that Bay issues, with the exception of wetlands protection, were eluding the grasp of the U.S. political and legal system. He observed that the conference participants posed these questions: How many people do we wish to house on the shores? How many tons of food do we wish to harvest, and what kinds? How big a ship do we wish to accommodate (implications for dredging and dredge spoil disposal)? How many pleasure boats will be operating? How many acres of wetlands should we preserve?

We have not yet faced up to any of those questions (with the exception of wetlands, soon after protected by law), and are only beginning to recognize them as legitimate issues. How big? How many? How much? Those questions seem almost to paralyze us, so directly do they suggest limits to our pursuit of the best of both worlds.[73]

The 1968 conference was followed in 1969 by the formation of an intrastate planning body made up of the heads of most of the state agencies having responsibilities affecting the Bay. This body, whose function was made largely obsolete by the major restructuring of state government in 1969 and 1970, was known as the Chesapeake Bay Interagency Planning Committee, and its activities resulted in the publication of a planning document.[74] In Virginia, similar focus was being placed on the James River, where at least two studies and numerous shorter publications of the Virginia Institute of Marine Science attempted to state the resource problems of the river, the relations between various uses, and the research and management needs posed by the increasing complexity and volume of use.[75]

Seen from the perspective of nearly two decades, these planning and study activities seem complex and intertwined. When considered in relation to the large number of important policy issues and controversies that were being addressed during that same period, they are nothing short of mind boggling. It is doubtful that they influenced legislation, public attitudes, and management practices that have since been adopted.

Not only did the decade of the 1960s see a vast increase in peoples's cognizance of the size and complexity of water-quality issues; it also saw the dividing line between water- and land-use issues blur and occasionally disappear. The relation between land activities and water quality took several forms. Most directly, this period saw the general acceptance of, and first action on, the role of non–point source pollution on the quality of the Bay. Second, it saw some recognition that general regional development had direct impacts on water quality. Third, this era produced a number of controversial projects that were opposed on water-quality grounds, but at root were conflicts over land use. And, finally, there developed a special concern for the role of wetlands in the physical and biological quality of the Bay.

That land runoff was a potential pollutant had long been recognized in the Bay region. Sediment was acknowledged as a cause of the reduced run of anadromous fish well before the twentieth century. Runoff from farmland was identified as a potential source of bacterial pollution by the Baltimore Sewerage Commission in 1897. The silting over of oyster bars as a result of heavy soil erosion was at least mentioned in the nineteenth century, if not greatly worried about. But it was after 1960 that government officials and institutions began to address the issue directly, beginning with soil erosion in Maryland.

Maryland state officials had discussed the impact of soil erosion and sedimentation prior to 1960,[76] and the Water Pollution Control Commission had concentrated on a specific aspect of soil erosion—that arising from sand and gravel operations—since the late 1940s. Special attention was given to both these problems, focusing on the Patuxent watershed. The commission began to worry about the heavy development of housing in the Bowie area, where it was announced that a large new community was to be developed in a previously agricultural area. This site was also on the same stretch of river that contained a number of sand and gravel operations and was not very far from tidewater, where there had long been concern for the gradually diminishing depth of water in the upper estuary. By the mid-1960s, with substantial interest from a number of legislators and local officials, the state developed a sediment control act for the Patuxent River watershed.[77] This act required local adoption of soil containment practices for large construction projects, backed by state enforcement powers and technical assistance. This program was adopted and successfully implemented on the Patuxent and was extended to the entire state in 1969.[78]

Both Maryland and Virginia also faced a relatively new problem during this period with the advent on a large scale of livestock herds and feedlots near tidewaters. This enterprise was considerably more of a problem in Virginia, where the concentration of hogs and cattle was higher. Large-scale feedlot operations were, in general, required to contain, if not treat, the waste from their operations, but extensive areas in Virginia were closed to shellfish harvesting at least partly as the result of agricultural land runoff. One of the difficulties in dealing with this problem was the uncertainty as to the causes of bacterial contamination. In a rural area there might be several sources contributing to excessive fecal coliforms: the discharges from the nearest sewage treatment plant; discharges from septic tanks; bacteria from wild animals, such as muskrats and geese; farm animals; and boats, particularly when concentrated in marinas. The difficulty of pinpointing sources, particularly over time, makes it difficult, both politically and technically, to require remedial action by one sector, even though it might be the most suspect. Runoff in the rural areas of Virginia thus required the closure of considerable areas to shellfish harvests.

Few sections of the country experienced more rapid growth than the Washington metropolitan area in the 1960s. One consequence was an attempt by local governments to control both the rate, character, and environmental impact of that growth. A variety of means were available: conventional zoning, performance standards for new subdivision construction, and extension of public services, particularly roads, water, and sewers. In addition, the unintentional but pervasive sewer moratoria that prevailed had the effect of curtailing or redirecting growth. Many of the reasons for growth management extended well beyond environmental concerns, but protection of streams and enhancement of the Potomac estuary were also recognized as purposes for which storm water management, sediment control, and open space protection were justified. Although the specific programs adopted by the several major jurisdictions are too numerous and complex to itemize, it is especially pertinent to note that local jurisdictions were recognizing that the heavy costs they were going to be required to bear in achieving advanced waste treatment of sewered wastes would perhaps be in vain if corresponding reductions were not also made in the waste loadings from non–point sources. This issue came under formal consideration on a regional basis under the aegis of the watershed planning provisions of the federal Water Pollution Control Act of 1972.[79]

It is a truism of environmental regulation that development projects are often opposed on environmental grounds when the actual reasons for the objection lie elsewhere. Proof of this in a particular instance is difficult, because it requires an inquiry into motivations of those who oppose a particular project. This difficulty is well illustrated by early experiences under the Maryland Wetlands Act: in case after case a tidewater construction project would be opposed under the act for reasons having nothing to do with the intent of the act. In general, opposition to small projects, such as bulkheads and piers, was based on disputes over riparian rights or questions of property boundaries. In many instances, opponents of a particular project did not even bother to address the wetlands protection issues involved.[80]

In the two cases that follow, the judgment that the opposition was arguing primarily over land use rather than water quality is that of the authors. Although the record is clear, at least, that the issues were mixed, these cases illustrate, notwithstanding, that non–water-quality issues became a factor in decisions that were reviewed or decided primarily on the basis of water-quality and related environmental considerations. Steuart Petroleum and Harry Lundeberg School of Seamanship, the two cases, physically and legally separate, took place in the same neighborhood at about the same time, and were opposed by essentially the same local interests and by the Chesapeake Bay Foundation. They illustrate a common principle: that local opposition based primarily on land-use issues can have a definite and perhaps decisive effect when expressed through government involvement in water-quality concerns.

In lower St. Mary's County, Steuart Petroleum operated an oil-handling and -shipping facility on the Potomac River, the primary function of which was to supply the Washington metropolitan area with petroleum products. Ocean-draft ships would unload at the facility, and the oil or other petroleum products would be transferred to barges or other shallow-draft craft for transport up the Potomac. In the late 1960s, Steuart announced its intention to apply for status as a free port, which would allow it to import foreign crude oil duty free. What was an oil-handling facility would become a refinery.[81]

Local opposition was swift and intense. A Potomac River protective association was formed and focused on the requirement of a building permit from the county commissioners before work could proceed. Opponents concentrated on the commissioners and alleged that the tax advantages of a major new facility, which were a prime attraction to a relatively poor rural government, would be more than offset by losses suffered by area watermen, recreation-oriented businesses, and property owners, due to the inevitable water-quality degradation that the oil facility would cause. Much was made of numerous oil spills that had occurred at the facility. Testimony from watermen argued that their catch had declined in the area due to oil pollution. Despite the company's claims that it would operate an environmentally safe facility (and despite the presence of a Coast Guard station in the shadow of the facility, thereby making it virtually inevitable that any violations of pollution statutes would be detected and prosecuted), the company was denied the necessary county permits. It thereafter withdrew its application for free trade port status and abandoned its plans. (Another refinery proposal was advanced by Steuart in 1973 and also was successfully opposed by local interests.)

The press labeled it an example of citizen concern for clean water.[82] It is arguable, however, that water was not the prime issue. The threat to local water quality was nebulous and unsupportable in advance of the construction. What was clear was that the facility would represent a major industrial intrusion, in this case an expansion, into an area notable for its isolation and natural charm.

On the other side of the narrow peninsula from the Steuart facility, St. George's Creek was being attacked by another intruder in the form of the Harry Lundeberg School of Seamanship. The school was operated by a small but politically influential union to train young people for maritime trades. In order to provide adequate draft and berthing space for the many vessels used by the school, the union embarked on a massive dredging campaign. Under permits from the Department of Water Resources, it operated its own hydraulic dredge, day and night, to dredge several hundred thousand yards of material, which it deposited within a large upland diked area across the creek from the school.[83]

Intense local opposition to the project alleged that it produced heavy sedimentation of oyster beds, that there had been oil spills from the dredge, and that the school was exceeding and breaking its permit restrictions in various ways. Neighbors also objected, understandably, to the night operations of the dredge, which made considerable noise. As a result of these complaints, various additional conditions were placed on the operation, and it was subject to several temporary shutdowns for violations.

In addition to the main dredging near the school, a state license had also been applied for to dredge a small cove on which the school owned other property, across and up the creek. With the passage of the Maryland Wetlands Act in 1970, a hearing on this project was held in May 1971. Local objectors to the project were numerous and passionate. They presented a litany of offenses and adverse consequences of the school's other dredging activities, claiming above all that the school was damaging the fisheries resources of the creek through siltation, destruction of the natural bottom, and pollution by oil and other wastes from the dredge and other vessels. Investigation by various technical specialists from the Department of Natural Resources suggested that these claims were probably not well founded, and that if the creek was undergoing environmental change it was due to a large number of factors, such as development in the watershed and along the shoreline, which were having a greater cumulative effect on the creek than would the relatively localized dredging by the school. Nonetheless, the license was denied, on the basis that the Wetlands Act established a presumption that undisturbed wetlands (in this case submerged lands) were in the public interest, and anyone wishing to alter or destroy wetlands had to make a case that there were offsetting benefits to be derived from the project. In this instance, the school was found not to have made a sufficiently strong case that the project was beneficial. Indeed, it appeared to the hearing

Marshland near Cove Point, Maryland, 1972. Preservation of sites similar to this has become a legislative issue. Since the photograph was taken, the Columbia Gas Company occupied the area for its liquefied natural gas facility. Photograph by Richard I. McLean.

officer that the school's primary objective was to convert a shallow cove into a bulkheaded harbor simply out of a seaman's preference for deep water and a firm shoreline.

What emerged from the testimony of many opponents of the project was that a primary concern was the very presence of the school itself. The school, like Steuart, was an intruder that threatened to alter the character of the area for retirees and longtime residents who enjoyed its isolation and natural beauty. The school portended continuing physical changes to the area and brought large numbers of strangers into an area that had been socially insular and stable. Certainly there were water-quality issues involved with both projects, but the overriding concern was with the more fundamental social, economic, and aesthetic changes that these two neighboring facilities represented. The water-quality and wetlands statutes administered by the state simply provided the most visible means for opposing them.

WETLANDS PROTECTION

Concern for wetlands protection became a national environmental issue in the 1960s. Numerous eastern states were enacting wetlands protection laws in the late 1960s and early 1970s. On the Bay, with its extensive tidal marshlands and shallow submerged lands, Maryland enacted a law to protect tidal wetlands in 1970[84] and Virginia did so in 1972.[85] The federal government weighed in with the Water Pollution Control Act of 1972 that closely regulates filling of wetlands.

Wetlands, in the form of swamps, bogs, and marshes, were once seen as little more than unsightly nuisance areas that bred disease and were better drained, filled, and put to new uses. That was the fate of much of the Bay watershed's wetlands: More than half have been lost since colonial times. That trend has slowed since wetlands came into their present status as "the heart of the Bay,"[86] providing habitat for waterfowl, wildlife, fish, and shellfish; buffers against erosion and tidal flooding; traps for sediment; and sources of beneficial organic matter and nutrients for aquatic life. In the last two categories, particularly, wetlands were seen as having a beneficial role in maintaining high water quality. (In Maryland a major impetus to the 1970 law was the prevention of land developers' enriching themselves by creating new land out of the public domain by dredging and filling. Here also, wetlands projects had land-use implications that were often of more public concern than their impact on the aquatic environment.)

Protection of both tidal and nontidal wetlands has since become an important component of the campaign to "Save the Bay." Wetlands can filter out as much as 85 percent of any phosphorus and nitrogen from agricultural runoff that flows through them. That makes wetlands essential to achieving the Chesapeake Bay Program goal of reducing the discharges of nitrogen and phosphorus into the Bay by 40 percent by the year 2000. In 1988, the Chesapeake Bay Program set an immediate goal of "no net loss" of wetlands and a long-term goal of recovering for a net gain some of those wetland areas that had been lost. The Fish and Wildlife Service has estimated that wetlands were destroyed at a rate of 4,500 acres a year in the Chesapeake Bay watershed between 1982 and 1989. By 1989, about 1.7 million acres of wetlands remained in the watershed. (Of that area, 12 percent consisted of tidal, saltwater wetlands along the shores and tidal tributaries of the Bay, and 88 percent was nontidal freshwater wetlands. Forested wetlands are the main wetlands form found in the Chesapeake, covering almost a million acres.) Confronted by these trends, Maryland, Virginia, and Pennsylvania acted between 1989 and 1991 to establish programs to regulate development and any other alteration of nontidal wetlands.

Enforcement of these and other state and federal wetlands protection regimes has been strict. Civil and criminal fines, imprisonment, and injunctions to restore altered wetlands await violators. Among a number of persons prosecuted for wetlands crimes, William Ellen, a Virginia engineer, was sentenced in federal court in Maryland in 1992 to a six-month prison term for violation of federal wetlands regulations. He had filled in wetlands while overseeing construction of a hunting preserve in Dorchester County.[87] In 1996, a Virginia developer, James J. Wilson, was sentenced to twenty-one months in prison and fined $1 million for directing employees to fill 70 acres of wetlands for a 9,100-acre planned community in Charles County. The two development companies he controlled were fined a record $3 million amid protests that such levies would force his companies into bankruptcy.[88]

Such penalties have been met with outrage in some quarters. Ellen became a martyr in a campaign of political conservatives and property-rights activists resisting regulators whom they label "eco-fascists." Margaret Ann Reigle, chairwoman of the Fairness to Land Owners Committee, a property-rights group based in Cambridge, defended wetlands violators as mere "dirt movers."[89] The more recent sentencing of Wilson in Charles County brought forth the inevitable comparisons to the fate of those who commit worse crimes. "I think it's so stupid that we attempt to incarcerate these people for moving dirt when heaven only knows what's going to happen to that environmental wacko, the Unabomber, who actually killed people," Reigle said.[90]

The regulatory scheme continues to expand. Anyone wishing to build on a tidal wetland in Maryland or Virginia may need both federal and state permits and must also comply with municipal and county ordinances. Wetland permits or license applications are granted only if (1) there are no practicable alternatives to the affected wetland—usually this means the proposed building or activity is dependent on proximity and access to the water; (2) changes to the wetland are avoided or minimized as much as possible; and (3) any alteration of the wetland is compensated for by the restoration of previously lost wetlands or the creation of new ones.

What seems a very fine regulatory filter, however, has proven more porous in practice than its proponents had hoped. More than 90 percent of applications for permits for activities that will disturb wetlands in the Bay watershed are granted. By the 1990s, enough development was slipping past regulatory scrutiny to thwart the "no net loss" goal for wetland preservation. In a 1994 report, the Chesapeake Bay Foundation found that wetlands permission programs "do not overburden development activity." Wetlands regulation was stemming the loss of wetlands, but not reversing it. Wetlands were still being destroyed illegally and legally through exemptions from the permitting process for projects involving isolated wetlands and for certain agricultural and forestry uses. Even those projects following the permission scheme did not always adequately fulfill the requirement to compensate for wetlands affected or destroyed. Sometimes wetlands replacement was never required as a condition for the permit. Other times, the compensation failed to replace the function and value of the particular area that was lost. And too often, compliance was unmonitored and unenforced.[91]

Among the recommendations made by the Chesapeake Bay Foundation to stiffen the permitting process were more funding for enforcement; new procedures to "encourage public involvement in permit review and enforcement"; the narrowing of certain general permit provisions and the exemptions for agriculture and forestry; and the requirement that compensation for affected wetlands be "full, in-kind replacement of the acreage and functions of wetlands destroyed."[92] Even these changes, however, would only "retard the loss of wetland acreage," the report concluded. To reverse the trend, new initiatives to restore wetlands would be needed. A perhaps unintended conclusion to be drawn from the report was that each new regulatory sally, while showing some success, seemed to reveal problems larger than anticipated, resulting in demands for yet more comprehensive controls.

The Chesapeake Bay Foundation's glum assessment of wetlands preservation notwithstanding, the Maryland regulatory framework erected in 1970 has led to a change in thinking about everything that could possibly flow, run off, leach, or fall into the Bay. Although as a statutory and administrative framework the Maryland law proved inadequate to handle the complexities of issues like Calvert Cliffs and Hart-Miller islands, it at least represented a start in bringing together for public consideration the range of water-quality and related environmental issues that the controversies of this era demanded. Clearly what was needed was not a static comprehensive plan for the Bay, which had been a popular cry of the late 1960s, but rather a process whereby the various and often competing interests of the Bay could be evaluated and balanced, based on some broad but reasonably explicit public policies. In virtually all cases, the starting question had been and will continue to be, What effect will this project have on the water quality and living resources of the Chesapeake Bay?

12

Acts of God and Acts of Man

The Chesapeake Bay is a national treasure and a resource of worldwide significance. . . . Man's use and abuse of its bounty, however, together with the continued growth and development of population in its watershed, have taken a toll on the Bay system. In recent decades, the Bay has suffered serious declines in quality and productivity.

—Chesapeake Bay Agreement, 1987

With the growth of the environmental movement, since at least the first Earth Day in 1970, the public at large has become more aware of environmental hazards, from the sensational oil spill in California's Santa Barbara Channel in 1969 to discoveries of pollution afflicting fields and streams around their own neighborhoods. In the 1970s, two such episodes—one an act of God, the other an act of man—jarred public awareness of the Bay's vulnerabilities while also giving testimony to its resilience. In 1972, tropical storm Agnes lashed the Bay watershed with record rainfall, sending torrents of freshwater runoff, sediment, and nutrients into the Bay. In 1975, revelations about the release of the toxic chemical Kepone into the James River seemed to indicate a threat to fish, shellfish, and wildlife of the Bay as well as the public health.

The dawning of this era of heightened environmental awareness brought a sense of alarm at every new discovery of ills besetting the Bay. Studies turned out more dire findings, and governments enacted more regulations in response. Political activist rhetoric in some quarters verged on the apocalyptic. The health of the Bay's water, vegetation, fish, and fowl showed signs of recovery in places; but suburban settlement continued its sprawl, and traditional Bay-faring communities declined as new protections for threatened species choked off their livelihood. New approaches sometimes revealed new problems calling for yet more sophisticated responses.

The Bay and its troubles were uncovered at deeper and more perplexing levels of complexity.

KEPONE

The Kepone chemical spill into the James River was one of the earliest environmental incidents to capture national attention[1] and was regarded as "the worst environmental disaster in Virginia's history."[2] Kepone, the registered trade name for the pesticide chlordane, is a powdery, gray-white, chlorinated hydrocarbon pesticide chemically related to DDT. It was used to dust potato crops in Europe and bananas in South America. In the United States, Allied Chemical Corp. (later Allied Corp., then, through a merger, Allied-Signal, Inc.) manufactured Kepone as a roach and fire-ant poison. Its toxic properties were well known: Initial tests by Allied, which owned the patent, indicated that Kepone caused cancer, liver damage, reproductive sterility, and inhibition of the growth and muscular coordination of fish, birds, and mammals.

Hopewell, Virginia, on the banks of the James River 70 miles upstream from the Chesapeake Bay, was the city where Allied manufactured Kepone from 1966 until 1974. In late 1973, the company granted an exclusive licensing agreement to Life Science Products, which had been started by two former Allied employees, to take over the manufacturing and sell the product to Allied. Hopewell played the delighted host. A welcome sign at the entrance to the city bore the slogan "Chemical Capital of the South."

Life Science began production of Kepone in March 1974 in fly-by-night fashion, manufacturing the chemical in an abandoned gas station and taking shortcuts from Allied's detailed production manual. Workers labored amid Kepone dust without protection of the gloves, boots, or respirators prescribed in the manual. Sometimes they picked up spilled Kepone with their bare hands. The fruits of such carelessness became evident in 1975 when a Life Science worker visited a physician, complaining of uncontrollable shaking. Kepone turned up in his blood and urine samples. After receiving notification of the test results, Virginia's state epidemiologist inspected the Hopewell plant on July 23, 1975, and found that seven employees there required immediate hospitalization. The next day, the state closed the plant. More worrisome details began to spill forth. Further tests revealed the presence of Kepone in more than half of Life Science's 110 employees. More than thirty of them were sent to the hospital for symptoms that included nervousness, memory loss, inability to concentrate, weight loss, liver damage, tremors, erratic eye movement, chest and joint pains, and sterility owing to reduced sperm counts. Fortunately, physicians treating the poisoned employees found that chlosetyramine, a drug used to lower blood cholesterol, was successful in removing Kepone from their systems. Follow-up tests showed an abatement of symptoms in many of the victims.

Of even greater concern than the effects on individual employees was the prospect of a threat to the general public health when the EPA discovered in 1975 that approximately 200,000 pounds of Kepone from the Allied and Life Science plants had been discharged untreated, and without authorization, into the James River—much of the spillage occurring while Allied was manufacturing the chemical. When Life Science took over, it initially sent wastes to the Hopewell sewage treatment plant, which imposed a pretreatment standard limiting the level of Kepone in the wastes. Even so, the sewage treatment plant was incapable of removing whatever Kepone remained in the wastes it received. Life Science's inadequate disposal methods came to the attention of state authorities when the company's Kepone-laden wastes disabled the sewage treatment plant's bacterial digester system in October 1974. The state and EPA responded by imposing a stricter pretreatment standard on Life Science, but the company's wastes continued to carry excessive amounts of Kepone.

The chemical was turning up on land, in the water, and in the air. Not easily dissolved by water, Kepone was settling into the sediment downriver from Hopewell. It was also detected in fish, oysters, and crabs downriver and upriver as far as Richmond, 20 miles away; in wildlife that fed on the James River's marine organisms; and in the atmosphere. The federal Food and Drug Administration established a maximum permissible concentration of Kepone for commercially sold finfish that effectively kept James River fish off the market in 1975 and 1976. At the same time, Virginia Governor Mills Godwin exercised his emergency powers to ban recreational and commercial fishing, crabbing, and oystering along a 98-mile stretch of the river from Richmond to the Hampton Roads Bridge-Tunnel.

Some scientists feared Kepone was reaching the Chesapeake Bay, where it would threaten commercial fishing and shellfishing. Although neither Maryland nor Virginia found the threat sufficient to restrict fishing there, many consumers at least temporarily avoided fish and shellfish taken from the Bay, and an economically devastating stigma attached. In Shady Side, a traditional Maryland fishing port, a commercial buyer found herself stuck with 8½ tons of bluefish and a market that wanted no part of her offer of 10 cents a pound. Not even dog- and cat-food makers were biting. In the end, she sold her stock for fertilizer at 1½ cents a pound.[3] A Richmond inn sought to assure panicked diners with a sign that said: "Fresh ocean oysters. No Kepone." In fishing hamlets such as Rescue, Virginia, watermen frustrated by what they thought was an overblown health scare ruining their livelihood slapped "Kepone Truckin' " bumper stickers on their pickups. Many reverently memorized a bit of data circulated by the Virginia Seafood Council that read: "A 132-pound person would have to consume 24,000 pounds or 72,000 meals of fish a year to be exposed to the amount of Kepone equivalent to that which caused harm in animals."[4]

Gradually, the crisis eased as the Kepone-laced sediment in the James was diluted and buried by uncontaminated sediment. In response to decreasing Kepone residues in fish samples, Virginia lifted its ban on the fishery in piecemeal fashion from 1975 to 1988. By 1978, EPA officials reported that the Kepone was unlikely to slide far downstream and that the Bay was not imminently threatened. But commercial fishing sustained lasting damage. In 1981, the Virginia Seafood Products Commission estimated that the state's then five-year-old ban on commercial fishing in the James had cost the fishing industry as much as $20 million and had forced some sixteen hundred Virginia watermen out of business.

Meanwhile, there was justice to be done. Kepone was removed from commerce. A few weeks after closing the Life Science plant in Hopewell, the EPA ordered the company to cease all manufacture and sales of Kepone. A year later, with the consent of Allied, the EPA canceled federal registrations for Kepone-based ant and roach poisons, thus making the sale of the product illegal in the United States. Allied and Life Science were forced to pay for their damage. The two companies and some of their executives faced extensive civil and criminal liability. Allied ended up pleading *nolo contendere* in August 1976 to 940 counts in a criminal charge of polluting the James River—225 of them stemming from single-day discharges of Kepone without a federal permit. A federal judge later acquitted Allied of all counts in two other indictments in which the company had been charged with conspiracy and with aiding and abetting Life Science's illegal discharges. The *nolo contendere* plea was designed to help Allied in its civil litigation; unlike a plea of guilty or a criminal conviction, a plea of *nolo contendere* cannot be admitted into evidence in subsequent civil suits for damages.

At the October 1976 sentencing on the 940 counts to which Allied had entered its plea, Judge Robert Merhige of the U.S. District Court in Richmond imposed a criminal fine of $13.24 million that he said was intended to send a message to corporate polluters generally. "I hope after this sentence, that every corporate official, every corporate employee that has any reason to think that pollution is going on, will think, 'If I don't do anything about it now, I am apt to be out of a job tomorrow.' I want the officials to be concerned when they see it."

The judge's ringing words got through to some management suites. The sentencing "was a shock to everybody and brought a real awareness, because we saw . . . thereafter that other companies were finally taking notice," said Manning Gasch, Jr., who represented Allied in the civil litigation.

After handing down what was then the largest criminal pollution fine ever imposed by a federal court, Judge Merhige was nevertheless "quite upset" that all that money would be going to the federal government. He believed it would be better spent directly on helping Virginia's environment.[5] At a hearing in February 1977, the judge reduced the fine to $5 million in exchange for Allied's agreement to pay $8 million to the Virginia Environmental Endowment,[6] a group recently established by Allied to support research into Kepone's effects, methods of removing it, and ways to improve the environment generally.[7] As part of the deal, Allied conferred authority to appoint the endowment's directors upon the senior federal district judge of the Richmond Division of the Eastern District of Virginia. That judge happened to be Robert Merhige, who continued to exercise that authority into the 1990s.

The benefit to Allied was to claim the $8 million as a tax-deductible business expense. The U.S. Justice Department strongly objected to the establishment of the fund because of the tax break it might give to Allied, but Judge Merhige said Allied had a clean record before the spill and should be commended for establishing such a fund. The company managers were "good boys in my book," he said.[8] The Internal Revenue Service didn't see it that way and denied the deduction. The U.S. Tax Court upheld the IRS decision in 1992, finding $8 million to be nondeductible because it was a penalty imposed "for punishment and deterrence of environmental crimes."[9]

This was just the beginning of Allied's liability. The company paid out millions in civil claims: $5.25 million to the city of Hopewell as part of a settlement;[10] more than $15 million in settlement of claims by fifty-six Life Science employees, their family members, and neighbors of the plant; and an undisclosed settlement in 1982 of a class action suit brought by commercial fishermen, oyster bed lessees, and bait and tackle shop owners claiming economic loss owing to Kepone pollution. (Judge Merhige, however, dismissed the claims of seafood wholesalers, distributors, and processors as well as restaurant owners and their employees for lost profits.[11]) In all, the Kepone debacle cost Allied at least $200 million in criminal fines, settlements, and legal fees.

The James River and the Bay emerged without serious long-term injury and with the protection of new regulations. Symptoms suffered by the Kepone poisoning victims have largely disappeared or improved. Amendments to federal and state environmental laws over the last twenty-five years have dealt with the lax regulation that made the Kepone spill possible. According to Professor William Goldfarb of Rutgers University, "It seems clear that given the current level of public attention, public resources and institutional coordination and commitment devoted to the preservation of the Chesapeake Bay, continuing pollution on the massive scale of the Kepone incident would be inconceivable. For example, the appalling lack of coordination among Federal agencies, state and local governments, and local citizens was one of the major reasons why the Kepone pollution went virtually unaddressed by regulatory authorities for nearly a decade."[12] Additionally, new technology has emerged to police discharges. A program initiated by Professor Robert J. Huggett of William and Mary College can now detect hundreds of compounds, including toxicants, in a single environmental sample. Virginia uses this so-called fingerprint method to identify and curtail toxicants in effluents that neither the state nor the waste generator had previously suspected.

The subsidence of Kepone's effects nevertheless leaves a vestigial stigma. Virginia fishing licenses still bear a notice to recreational fishermen that there is Kepone in the James River and that it may be hazardous to their health. And the city of Hopewell no longer bills itself as the "Chemical Capital of the South."

AGNES AND OTHER STORMS

The forces of nature can ravage the Bay worse than man-made pollution. For three days in June 1972, tropical storm Agnes poured record amounts of rain into the Chesapeake. All parts of the Bay watershed were soaked by at least 5 inches of rain, with one-third of it receiving more than 12 inches, and some areas enduring as much as 18 inches. The resulting floods stripped wastes from the region and flung them into the Bay, leaving it a roiling mess awash in bacteria, sewage from ruptured lines, sediment, organic matter, farm chemicals, and debris ranging in size from plastic bottles to entire houses. Parts of the Bay were freshened past tolerance for saltwater plants and animals. In Agnes's wake was a Bay fouled, corrupted, and in the worst condition of its recorded history.[13]

Beholding the destruction, author Annie Dillard observed:

It's all I can do to stand. I feel dizzy, drawn, mauled. Below me the floodwater roils to a violent froth that looks like dirty lace that continuously explodes before my eyes. If I look away, the earth moves backwards, rises and swells, from the fixing of my eyes at one spot against the motion of the flood. All the familiar land looks as though it were not solid and real at all, but painted on a scroll like a backdrop, and that unrolled scroll has been shaken, so the earth sways and the air roars.

Everything imaginable is zipping by, almost too fast to see. If I stand on the bridge and look downstream, I feel as though I am looking up the business end of an avalanche.[14]

The Susquehanna River, which in a normal year discharges 500,000 to 1 million metric tons of sediment into the Bay, disgorged 31 million tons in only ten days. In the sense that the Bay is a basin that will cease to exist when it is finally filled with sediment, this inundation by Agnes "aged" the Bay in some places "by more than a decade in a week."[15] Nitrogen concentrations were increased by two or three times in the upper half of the Bay, causing abnormally large algae blooms. The blooms and the sediment, up to a hundred times normal levels in some parts, blocked sunlight, thereby killing about half the Bay's bottom grasses. Shellfish, deprived of the requisite salinity in their waters, also suffered. More than 90 percent of market-size soft-shell crabs and about 2 million bushels of market-size oysters died as a result of the storm.

Some commentators were reluctant to lay all of the blame at the feet of nature. "Agnes would have been a shock to the Chesapeake's system in any age, but, coming in modern times, it almost surely was far more devastating," wrote *Baltimore Sun* environmental reporter Tom Horton in his 1987 book, *Bay Country.* "The truth may be that Agnes wasn't the problem so much as the incapacity of the modern bay—shorn of its biological filters and buffers, destabilized and stressed already to the limit—to handle the insult."

On a smaller scale, each seasonal storm seems to have some deleterious effect on the Bay. The large amounts of rain in the Bay watershed in September 1996 from tropical storm Fran may be found to have caused harm similar to Agnes. Extended rains and heavy snow melt also send down damaging flows of fresh water. The freshets following the harsh winters of 1993 and 1994 lowered salinity intolerably for many oysters, returned an algae bloom to the Potomac, reduced oxygen levels in bottom waters, and, after years of steady resurgence from the havoc of Agnes, slowed the spread of Bay grasses.[16] Even mild winters do harm. The snow melt was so slight from the winter of 1995 as to break the record for low water flow from the Susquehanna. At only 35,700 cubic feet per second, the flow was not strong enough to push the algae bloom resulting from nutrients in the spring freshet south of the Chesapeake Bay Bridges. Consequently, the deep channel north of the bridge suffered low oxygen levels while the waters to the south had higher levels than usual.

The effects of that mild winter, compared with those of severe ones, showed that the actual health of the Bay in a given year depends primarily on the amount and timing of freshwater flows, all of which raises the question of what would be an ideal winter for the Bay, or whether such a phenomenon is possible. In that respect, according to a National Oceanic and Atmospheric Administration official in the Chesapeake Bay office, "Every year is special."[17]

HOW ARE WE DOING? YET MORE STUDIES

The Chesapeake Bay is perhaps the most studied estuary in the world, yet every finding poses more questions and every fresh question begs for answers. Some studies confirm each other as to the perilous condition of the Bay. Others indicate life is improving. Some prompt government action, others gather dust. A quick review of studies of the Bay's health suggests some problems have abated,

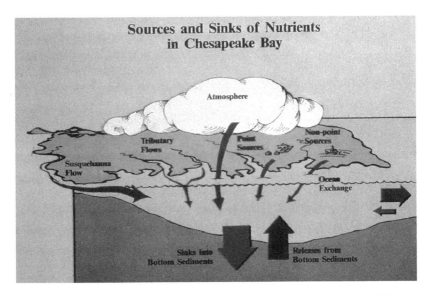

The nutrients phosphorus and nitrogen, which cause harmful algae blooms in the Bay, enter the Bay from farm runoff, deposition from air pollution, and direct discharges from sewage treatment plants. Courtesy the Alliance for the Chesapeake Bay.

while some are getting worse, and new questions are poking through.

The 1977 Bi-State Conference on the Chesapeake Bay, held at the Patuxent Naval Air Station in Lexington Park, Maryland, reported "excessive algae blooms" that were derived from "large amounts of nitrogen and phosphorus" coming into the Bay, largely from manure and the fertilizers in farmland runoff. The result was a diminution of submerged plant life. Yet, whether related to this trend or not, bluefish proliferated while shad and herring were scarce and striped bass were on the way down. The conference also found more low-oxygen zones and "massive accumulations" of metals, from domestic and industrial wastes, lodged in harbor sediments.

These findings did not lead to significant government action, whereas the EPA's subsequent Chesapeake Bay Study, confirming them, did. The study produced five reports released in 1982 and 1983, at a cost of $27 million, concluding that the estuarine environment was declining. Algae blooms were spreading; bottom grasses were shriveling, both in abundance and variety; marine spawners such as menhaden and bluefish were on the rise, while freshwater spawners such as striped bass were decreasing; oyster harvests had dropped as had their rates of reproduction; and many ducks, particularly pintails, widgeons, and redheads, had departed the Bay for lack of grasses. In something of a breakthrough, the study showed that, although sewage treatment plants contributed to the phosphorus and nitrogen in the Bay, "the fact that agricultural practices are a significant contributor was conclusively shown for the first time."[18] The study also reported that toxic organic compounds and metals turned up in high concentrations near industrial sites at Baltimore and Norfolk, though damage by these pollutants was confined to nearby ecosystems.

The study proved a triumph for environmental activists. Nearly gloating in his annual report, Chesapeake Bay Foundation President William C. Baker recently wrote that the study "put to rest the whining excuses of anti-environment critics who had said the Bay's declining state was the result of 'natural cycles.' " Unlike preceding studies that produced many of the same findings, this one led to a commitment from the EPA and the four states bordering the Bay to work together toward more comprehensive approaches to problems.[19]

Studies into the effects of air pollution on the Bay are less conclusive. The 1983 Chesapeake Bay Study never identified air pollution as a significant problem in the Bay, but more recent research has found that 25 percent to 35 percent of nitrogen pollution there comes from the emissions of automobiles and industrial smokestacks. Through computer modeling, EPA scientist Robin L. Dennis traced two-thirds of the nitrogen entering the Bay watershed to sources beyond the control of Bay watershed governments—Ohio River power plants and industries and automobiles, many of which may be passing through from out of state.[20] With respect to boats on the Bay at least, new EPA regulations under the Clean Air Act will require significant reductions in emissions of hydrocarbons and nitrogen oxides from new gas-and diesel-powered marine engines.

Air pollution is the latest realm of scientific inquiry and speculation into the Bay's troubles. A 1994 EPA report called it "the single biggest source of toxic chemicals in the Bay, with storm-water runoff the second largest source," displacing the toxic troika of factories, military bases, and sewage treatment plants.[21] This ranking remains open to question, however, because of the uncertain and not easily measured routes that pollution may take from air to land to water. Pollutants arrive mixed with precipitation or fall directly to earth. Those that fall on land instead of water may reach the Bay in water runoff or, after leaching, through groundwater. Along either route, they may break down into less toxic substances before reaching the Bay.

Pollutants dropping anywhere within the Bay's 64,000-square-mile watershed eventually may make their way to the Bay,[22] but in what amount, in what composition, and at what level of toxicity? "How much is transmitted through the watershed to the Bay is really open to wild speculation," said Joel Baker, a researcher at the University of Maryland's Chesapeake Biological Laboratory. "It's greater than zero but less than 100 percent."[23] The speculation leads some scientists to fear that the unknown may be worse than the known sources of pollution. As Assistant Professor Rebecca Dickhut of the Virginia Institute of Marine Science points out, pollutants may exist at lower concentrations in the air than in the wastewater of a particular discharge pipe, but air pollution has access to the entire Bay watershed in which it can be exchanged from air to land to water.[24]

Newer studies not only uncover new problems but reinterpret old ones and question long-standing approaches to them. Scientists have argued for years over which is more harmful and thus more deserving of attention, nitrogen that supports algae growth in salt water, or phosphorus, which is the more important nutrient for it in fresh water. Lacking solid data, state and federal regulators have concentrated on phosphorus controls primarily because that is cheaper.[25]

Studies, however, have begun to challenge this common wisdom with the finding that nitrogen is more easily soluble in groundwater and thus flows more readily into the Bay. Moreover, a recent computer model study indicates that reducing phosphorus does little good without corresponding reductions in nitrogen. Robert Thomann,

Power plants (like Virginia Power's Yorktown plant above) and automobiles emit large amounts of nitrogen, which cause harmful algae blooms in the Bay. Courtesy the Alliance for the Chesapeake Bay.

a professor of environmental engineering at Manhattan College in New York, explains the theory: In the upper Bay, phosphorus stimulates the growth of freshwater algae, which absorb nitrogen. With less phosphorus, and consequently less freshwater algae to absorb nitrogen, there is more nitrogen flowing down to the saltwater portions of the lower Bay to stimulate algae growth there. In other words, solving the phosphorus problem worsens the nitrogen problem.

THE HARD PART: TURNING STUDY INTO ACTION

The recent history of Bay studies reveals afflictions so numerous, persistent, and often elusive of traditional government jurisdiction that public commitments to curb them, at considerable public expense, have yielded only mixed results. A review of the chief concerns of the 1990s—toxic pollutants, warming, sediments, erosion, rising water, and the decline of waterfowl and fisheries—illustrates the point.

Toxic Pollution

In the area of toxic pollution, one issue is defining the term and the other is deciding which of the more than one thousand different chemical substances detected in or discharged into the Bay watershed qualify under it. The federal Clean Water Act labels 126 naturally occurring metals and man-made chemicals as "toxic pollutants."[26] In 1991, the Chesapeake Bay Program listed fourteen "toxics of concern" representing an immediate or potential threat to the Bay. A 1994 report of the program found that the Bay ". . . does not appear to have a severe systemwide problem with toxics," but that some areas do have "severe problems" and that "some levels of toxics can be found almost anywhere" there.[27] Where the toxics abound, the effects can be glaring. In the highly polluted waters of the Elizabeth, Patapsco, and Anacostia rivers, fish suffer from liver tumors, cataracts, lesions, diminished respiratory ability, and high mortality rates. In the Chesapeake and Delaware Canal and certain rivers, striped bass yearlings have deformed gills. Birds, mammals, reptiles, and amphibians around the Bay, however, have so far been spared such harm.[28] Attacking such pollution requires fighting on all fronts: air, land, and water.

Amendments to the 1990 Clean Air Act were aimed at the arrival of pollutants by air. Although not intended to benefit the Bay, the legislation's mandate to lower emissions of nitrogen oxides from new automobiles and electric power plants may reduce by 17.5 percent the amount of nitrogen reaching the Chesapeake. The Bay also may benefit from other amendments to the act that seek to encourage development of low-emissions automobiles, cleaner fuels, and stricter automobile inspection programs to cut the ozone (otherwise known as smog) and carbon monoxide in polluted urban areas.[29] The question raised by such measures is whether they can remove pollution as fast as emissions increase from a greater number of automobiles being driven greater distances.

The political will was sufficient to move broadly against air pollution. In 1991 Maryland Governor William Donald Schaefer said that "air pollution from cars is a major threat to the Chesapeake Bay and will become a new target of the state's cleanup campaign."[30] Congress in 1990 ordered tougher vehicle emissions checks in eighty metropolitan areas around the country. Maryland was one of the first states to attempt to meet the federal clean air requirements with an inspection program for cars and light trucks. As the program was about to take effect in 1995, Marylanders' passions for their privacy and rights to their own cars were aroused, pushing the issue to the forefront of political debate over how much personal sacrifice citizens must make for the good of the environment. Car owners balked at the intrusiveness of having to let emissions test site workers get behind the wheel and "drive" their cars the equivalent of two miles on a treadmill at speeds of up to 55 mph. Strict standards all but guaranteed a 20 percent chance of flunking the test and having to return for reinspection after paying for a costly tune-up or other repairs. Radio talk show lines crackled with rebellion. "I can't tell you how outraged people are," said Martha Klima, a Republican state legislator who drafted a bill to repeal the 1991 General Assembly law authorizing the tougher testing. "I think we all want clean air, but there's got to be a balance between the rights of citizens and the mandates of government."[31]

Incoming Governor Parris Glendening found a way out—a moratorium passed by the General Assembly to put off the strict new regime for eighteen months while state officials evaluated the federal pollution guidelines and explored ways to reduce repair costs for cars that failed the test.[32] In Virginia, Governor George Allen successfully sued the federal government over its threat to withhold federal transportation funds if the state did not adopt a tighter emissions program.[33] Regarding other areas of the program, however, Maryland authorities had not backed down. "We have hit big sources [of smog] hard, and they are going to be hit again," explained David A. C. Carroll, the state's environment secretary. "Now we're picking on everybody. We're going all down the line."[34]

The main land-based sources of toxic pollutants going into the Bay are the 6,000 industrial sites, military bases, and sewage treatment plants along its shores. The National Oceanic and Atmospheric Administration estimates that about one-fifth of the water entering the Bay at any one time is wastewater from industry and sewage treatment. The Clean Water Act regime limits effluents, but not entirely. The Blue Plains sewage treatment plant in southwest Washington, D.C., believed to be the largest source of nitrogen pollution in the Bay, removes 99 percent of the phosphorus from the 309 million gallons it discharges daily into the Potomac. The plant stabilizes but does not remove nitrogen from the effluent. The 14 million pounds of it flowing into the river each year is an improvement over the past that has ushered in the return of striped bass and bottom grasses to the river. As the Bay area's population continues to grow, however, more sewage treatment plants will spring up, and their discharges, though heavily treated, will add to nitrogen pollution.

Runoff from farms and land development is a pollution source much harder than wastewater to contain. There are 24,000 farms in the Bay area that send manure (containing phosphorus), fertilizers (containing nitrogen), pesticides, eroded soil, and other wastes streaming into the Bay when rain falls and snow melts. The farm animals of Lancaster County, Pennsylvania, alone produce 30 million pounds of manure a day. The paving of land for roads, homes, and commercial and industrial uses destroys wetlands, trees, and other vegetation that perform the invaluable task of absorbing and filtering runoff pollutants. About 360,000 acres of Central Maryland have been covered for housing development since colonization under Lord Calvert. The Maryland Office of Planning projects more than that will be developed for housing in the next twenty-five years as population increases and residents abandon cities and established suburbs for new houses built on larger, dispersed lots inhabited by fewer persons per household.[35]

Runoff from a typical housing subdivision and the roads that serve it can bring sewer overflows, pesticides and fertilizers from lawns, sediment from construction sites, zinc and cyanides from road salt, chromium from paints and stains, cadmium from corroded alloys and motor oil, and copper from brake liners and pipe corrosion. A nontypical subdivision, the huge proposed Chapman's Landing development in Charles County, would convert 2,250 acres of forests into 4,600 housing units, 2.25 million square feet of commercial space, a golf course—and a source of sediment and nutrient runoff large enough to threaten the Potomac's productive Mattawoman Creek fishery. Development is hard to stop, despite the spread of more restrictive zoning. The issue is how it will continue, at what pace, in what density, in what form of trade-offs for the health of the Bay.

Not all runoff is inevitable with the coming of residential sprawl; small improvements are possible. For example, chemical runoff can be reduced if homeowners follow the instructions on the bag of lawn fertilizer. Studies conducted at the Chesapeake Bay Foundation's

Clagett Farms in Upper Marlboro, Maryland, in 1990 concluded that lawn chemicals applied in the prescribed amounts proved harmless to groundwater and the Bay. When twice the right amount was used, the excess ran off the soil. "The problem is many homeowners do not follow the guidelines," said J. Scott Angle, a University of Maryland agronomy professor who directed the studies. "They say, 'If I apply a certain amount of fertilizers to my lawn and it looks this good, then I'll apply twice as much and it will look twice as good.' But the lawn can only assimilate so much and the excess has to go somewhere; most of the time it's the Chesapeake Bay."[36]

Regulation, too, has had a measurable impact. State bans on phosphorus detergents in Maryland, Pennsylvania, Virginia, and the District of Columbia and improvements in sewage treatment helped lower phosphorus discharges into the Bay by 16 percent between 1985 and 1992. More recent data, however, suggest that nature is no respecter of man's best-laid plans to nurture the environment. Strong freshets in 1993 and 1994, along with flooding in 1996, appear to have returned phosphorus concentrations in the Bay to their 1984 level and increased nitrogen concentrations in the Bay 10 percent since that time. Despite the extra nutrients, however, algae concentration has dropped 18 percent in the Bay since 1984, probably because the most recent load of phosphorus comes bound to dirt, plants, or animals and thus is unavailable as a food source for algae. Dissolved, inorganic phosphorus, the form that appeals to algae, has stayed the same since 1984.

Warming of the Bay?

In 1991, the EPA reported that Chesapeake Bay waters had warmed by almost three degrees since 1985. Scientists feared the trend would decrease oxygen and render the Bay less hospitable to various forms of marine life and more inviting to marine organism diseases and earlier algae blooms. Some envisioned the demise of soft clams and fewer oysters, blue crabs, and striped bass.

But consensus has yet to form on the nature and extent of the threat. For instance, Chris D'Elia, head of the Sea Grant College at the University of Maryland, contends that algae blooms depend more on sunlight and the size of spring freshets than upon water temperature. Scientists also disagree as to the cause of the Bay's rising temperatures. Some attribute it to global warming—the theory of "greenhouse" gas emissions trapping the sun's heat reflected off the earth's surface—but the proposition of global warming itself is vigorously contested. Other scientists believe that natural temperature cycles, from milder winters and hotter summers, are the cause, and accordingly, the Bay's temperatures may decrease in the future.[37]

Sediment Pollution

Sediment entering the Bay displaces water, gradually filling it and disturbing life at the bottom. The sediment smothers vegetation and immobilizes shellfish. It also muddies the waters. When less sunlight pierces the murk, photosynthesis of submerged vegetation diminishes, hastening its decline.

The Army Corps of Engineers declared in a 1991 report[38] that 4.7 million cubic yards of sediment erode and wash into the Bay each year and another 4.3 million cubic yards flow in with runoff through its tributaries. More recent studies indicate that the sediment eroding directly into the Bay is loaded with nutrients, contributing 5 percent to 10 percent of phosphorus and nitrogen pollution.

In addition to destroying buildings built too close to the Bay and its tributaries, erosion causes harmful sediment to be added to the Bay. Courtesy the Alliance for the Chesapeake Bay.

Erosion along the Bay and Its Tributaries

Erosion is caused by wind pushing water toward land and by rising water levels. The corps' 1991 report estimated that the Bay and its tributaries lose an average of 1 foot of shoreline a year and that 45,000 acres of beaches, bluffs, and wetlands—an area almost the size of the District of Columbia—have tumbled into the water over the last one hundred years. A U.S. Fish and Wildlife Service study found that since colonial times, erosion has entirely wiped away twelve of thirty-five islands in the middle section of the Bay and along the Eastern Shore, for a total loss of 10,500 acres. Some of the loss is historic: The grave of George Washington's great-grandfather at Millenbeck Plantation in Lancaster County, Virginia, may have collapsed into the Corrotoman River, a tributary of the Rappahannock. And according to

This marsh planting is an example of a nonstructural erosion-control method, which helps lessen the impacts of erosion on water quality and habitat in the Bay and its tributaries. Nonstructural methods often have a structural component such as riprap to help protect the planted area. The Chesapeake Bay Program prefers nonstructural methods of erosion control because they retain transitional habitat between open water and upland and do not result in scouring of the sandy bottom, which may occur when waves strike a sheer seawall or bulkhead. Courtesy the Maryland Chesapeake Bay Critical Area Commission.

Randolph Turner, chief archaeologist of Virginia's Department of Historic Resources, hundreds more historic encampments, settlements, plantations, and forts may have joined the sediment of the Bay and its tributaries.

Efforts to prevent erosion raise what has become a typical dilemma of Bay management—solutions that provoke new problems. Shoreline property owners attempt variously to stabilize their land with "hard" control methods such as bulkheads and riprap revetments, and nonstructural measures such as grading and cultivating shore vegetation. But Duke University geologist Orrin Pilkey and others believe the hard structures cause more erosion elsewhere than they prevent, while altering the natural shoreline and adjacent habitats. Government authorities such as Maryland's Critical Area Commission have heeded Pilkey and discouraged or prohibited bulkheads and revetments in favor of nonstructural approaches.

Rising Water Levels

A key factor in erosion is rising water, a trend tied into many others besetting the Bay. Estimates of the rise range from 4 inches to 1 foot over the last century. Professor Michael Kearney, who teaches geography at the University of Maryland, puts his estimates at the high end of that range and predicts that the next century will bring an increase of 3 more feet, inundating 1,000 feet of Bay shoreline. Beaches, marshes, and some islands, such as Smith and Tangier, might disappear in the deluge. Species that rely upon marshes would lose habitat.

Believers in global warming foresee polar ice caps and glaciers melting, leading to a 3- to 5-foot rise in ocean waters within one hundred years. Whether global warming is occurring, however, is an unsettled question in the scientific community. For those who subscribe to natural temperature cycles, the future of water levels seems less dire.

Decline of Waterfowl and the Grasses That Feed Them

Most waterfowl like the Bay best in winter. Approximately one-third of the waterfowl wintering along the Atlantic Coast choose the Chesapeake. But species that used to feed on the roots and leaves of submerged vegetation have been particularly hard hit by the significant loss of the grass and other plants on the Bay's bottom and fly elsewhere to spend the winter. Tundra swans now go to North Carolina.

Canada geese, once the most abundant waterfowl on the Bay, are much scarcer now. In recent years, mallards have come back, but black ducks have been lost.[39]

The great attraction for waterfowl—submerged grasses—once covered over 600,000 acres of the Bay. The vegetation reached a low of 38,000 acres in 1984 but recovered to 73,000 acres by 1993. New vegetation such as hydrilla has improved water clarity and habitat for fish and fowl in the Potomac River; widgeon grass supports a blue crab population around Smith and Deal islands and Crisfield. The decade-long progress was brought up short in 1994 when grass-covered bottom dropped to 64,000 acres,[40] owing to freshets bringing large amounts of nutrients into the Bay.

Decline of Bay Fisheries

The seafood harvest of the Bay is not what it used to be, and measurably so. Overharvesting, disease, pollution, dams that block access to fish spawning grounds, and degradation of marine habitat all share the blame for the losses among the once teeming oysters, striped bass, American shad, clams, river herring, and blue crabs. Add to those culprits the U.S. and Maryland constitutions, which courts interpreted in 1971 as imbuing watermen with the right to cross the Bay's county boundaries in pursuit of shellfish. They reaped a harvest of diminishing returns, however. The result, according to Tom Horton in his book *Bay Country*, was "a more mobile work force, with bigger, bay-ranging boats that can now exert tremendous pressure within a matter of hours on any new 'hotspot' where the beleaguered oysters and other species try to stage a comeback."

During the summer of 1997, fishermen in various areas around the Chesapeake Bay, particularly in the Pocomoke River, reportedly caught striped bass, perch, and other fish exhibiting unexplained ulcerous sores, raising new concerns about the Bay's fisheries. At the time of this writing, scientists had not established the cause of these lesions, but a suspected cause was the one-cell microbe *Pfiesteria piscicida*, which was also believed to have been responsible for killing millions of fish in North Carolina.[41]

OYSTERS

Even greater than its worth as a commercial seafood catch is the oyster's invaluable talent for filtering algae and sediment as water passes through its gills while it is feeding. In the 1870s, when there were 12 million bushels of oysters in the Bay, they could collectively filter all its water within a week. Now, with oysters down to 1 percent of that population, they need almost a year to do the job. The obvious lesson is that more oysters boost the chances for cleaner water, but the trend has been downward in the long term, uneven in the short term. The peak harvest was 15 million bushels in 1885. The low was 7,000 bushels in Virginia and 70,000 in Maryland during the 1993–94 season. Crabs have since replaced oysters as the premier shellfish catch of the Bay. By the mid-1990s, the traditional St. Mary's Oyster Festival was celebrated with oysters from outside the county. The St. Mary's River harvest had shriveled to 34 bushels in 1993, from 71,000 bushels a decade earlier.

Much of the drop in recent years is attributable to two diseases, Dermo (*Perkins marinus*) and MSX (*Haplosporidium nelsoni*), which kill many oysters before they reach market size but allow them to live long enough to reproduce. The diseases, caused by protozoan parasites, abhor fresh water; the trouble is, so do oysters. Still, oysters generally come out ahead by attrition when the Bay is freshened. Strong spring freshets in 1993 and 1994 curtailed the instance of Dermo and MSX infection, but rates climbed back the following year.

As the destruction accelerated, environmental activists sounded alarms. State governments responded, but differently. In 1991, the Chesapeake Bay Foundation called for a three-year moratorium on taking oysters in the Bay. The idea was that the oysters might have a chance to develop immunity in the interim to Dermo and MSX. In what sounded to the thousands of watermen and seafood packers as an afterthought, the foundation proposed that those who make their livelihood in Bay oystering spend that time "in expanded programs to rebuild oyster populations."

Watermen were outraged. The moratorium idea was nothing but a "publicity stunt," said Larry Simns of the Maryland Watermen's Association. "Why don't they close the farmer for three years, or close industry for three years? They figure we're an easy target."

Simns was only half right. Virginia did ban oyster harvesting in 1994 and 1995 in all of the state's portion of the Bay and in most of its tributaries. Maryland refused. "I'm not convinced that [oysters] are an endangered species," said Maryland's Secretary of Natural Resources Torrey C. Brown, "and I'm not convinced that a ban would work for oysters." Maryland instead established 200,000 acres of sanctuary where oystering is banned and began spending $2 million a year to reseed historic oyster bars.[42]

STRIPED BASS

As went the Bay, so went the population and harvests of the striped bass. The Bay has historically spawned 90 percent of the striped bass on the Atlantic Coast. The drop in this population, locally known as rockfish, was precipitous: from a catch of 7.3 million pounds in 1970 to 2.5 million pounds in 1980 to 600,000 pounds in 1983. As a consequence, authorities enacted a five-year ban on catching striped bass, starting in January 1985. The measure was a complete success, and fishing gradually resumed. The number of young striped bass in 1996 reached a high for the forty-three years in which such records have been kept. In 1995, the Atlantic States Marine Fisheries Commission officially declared the species to have recovered in the Bay.[43]

AMERICAN SHAD

American shad was once the most bountiful fishery of the Bay, yielding as much as 2 million pounds a year before 1973. By 1980, their number had dwindled to about 5,000. From that year up to the present, Maryland has prohibited the taking, possession, or sale of American shad from the Bay. A similar ban took effect the following year on hickory shad. The District of Columbia followed suit in 1989, Virginia in 1994.

What caused the decline of the shad is something of a mystery. One theory is that bigger catches of shad in coastal ocean waters depleted spawning stocks in the Bay and its rivers, although some scientists contend the ocean fishery has little connection to the estuarine one. The devastation of the shad has also been attributed to the usual suspects: pollution and overharvesting, as well as to construction of more than seventy dams that block the anadromous shad from swimming upstream to freshwater spawning grounds.

Left alone by fishermen, shad have regrouped. In the upper Bay, shad reached an estimated population of 141,000 in 1991, dipped to 47,000 in 1993, but recovered to 337,000 in 1995. Tempering the good news, Richard St. Pierre of the U.S. Fish and Wildlife Service has attributed the comeback not to any improvement in habitat but to upriver hatchery releases.

Shad are now assisted in their upstream reproductive mission by fish lifts constructed at dams on fifteen rivers. The new fish passages on the James River will return American shad upstream to the Piedmont area after an absence of more than a hundred years.[44]

CLAMS

The decline of the clam has been swift, from a harvest of more than 365,000 bushels in 1989 to less than 20,000 in 1992. The intrusion of more fresh water and warmer temperatures are to blame.

RIVER HERRING

River herring—the alewife and blueback—were among the most plentiful in the commercial harvest of finfish from the Bay in the 1930s. By the 1970s, overfishing, pollution, and loss of spawning habitat had sharply reduced their numbers. The catch of river herring dropped more than 80 percent between 1965 and 1985.

BLUE CRABS

The Bay once produced more blue crabs than any other body of water in the world. After precipitous declines, it still supplies almost half the crab harvest in the United States. The blue crab's numbers remained robust until the late 1980s. By the early 1990s, the Gulf of Mexico surpassed the Bay's yield, which had dropped to 30 million pounds in 1992 from a previous average of 47 million pounds. Scientists offered competing theories to explain the change and differed over whether it was even substantial or permanent. Watermen and environmental activists disagreed to the point of insults and threats over whether crabs were overharvested.

The National Oceanic and Atmospheric Administration reported in early 1996 that blue crabs were not overharvested and that the decline was part of a natural cycle from which they could recover. Watermen flocked to this position. But other theories abound. One is that commercial harvests have been taking more female crabs in recent years. Another is that the scarcity of oysters has prompted watermen to fish for crabs longer.[45] A more intriguing possibility is that crabs are still recovering from tropical storm Agnes, which destroyed 90 percent of the lower Bay's eelgrass, a crucial habitat for juvenile crabs, and that subsequent high harvests kept the population low despite the recovery of submerged aquatic vegetation.[46]

Virginia and Maryland responded, beginning in 1994, with restrictions on the commercial and recreational taking of crabs from the Bay.[47] The Chesapeake Bay Foundation urged more dramatic action: a ban, to begin in September 1995, on harvesting crabs in all waters of the Bay 40 or more feet in depth—about a quarter of the Bay. Foundation President William C. Baker warned that the "crab fishery on the Chesapeake Bay is really on the verge of collapse."[48]

On Smith Island, watermen erected billboards denouncing the foundation for its support of state catch restrictions imposed in the fall of 1995. The signs said: "Smith Island's way of life will soon be over due to the Chesapeake Bay Foundation. Please do not support them." Concurrently, foundation property on the island was vandalized—including an outbuilding that was put to flame, though the culprit was not known.[49] The angry message behind the signs enjoyed support over much of the island but divided the village of Tylerton. Since 1978 the foundation had operated an education center there that annually drew about 2,500 middle and high school students for guided canoe excursions, and that helped sustain Tylerton's daily ferry service and its only store. When foundation officials wondered aloud how long the foundation could maintain its presence in the midst of such hostility, Tylerton's economic future suddenly became something to think about.[50]

Village pastor Ashley Maxwell, who tried to mediate the dispute, summed up the resentment: "The Chesapeake Bay Foundation is very rich and very powerful. I'm sorry to say that it's a matter of power." Some islanders see the foundation as a stepfather, he said, "coming in and saying, 'Who's got the money? If you don't play ball, I'm going to hurt you.'"[51]

Bill Goldsborough, a foundation scientist who had been the education center's first manager, explained, without making much headway among islanders, that the foundation had urged swift action on the crabbing issue so as to "act ahead of time to prevent a problem.... One of the main reasons that you want to save the bay is to maintain the watermen's culture."[52]

The blue crab battle brings into sharp relief the trade-offs between fisheries and livelihood, water quality and individual freedom, that have brought Bay politics from how fast government must act to clean up dangerous messes like Kepone to how far government should go before human economies and freedoms are impaired.

Managing an Integrated Ecosystem

*Representing the Federal government and the states which surround
the Chesapeake Bay, we acknowledge our stake in the resources of the
Bay and accept our share of responsibility for its current condition.
We are determined that this decline will be reversed. In response, all
of our jurisdictions have embarked on ambitious programs to protect
our shared resource and restore it to a more productive state.*
— Pledge by the governors of Virginia, Maryland, and
Pennsylvania, the mayor of Washington, D.C., the EPA
administrator, and the chairman of the Chesapeake Bay
Commission in the 1987 Chesapeake Bay Agreement

Every year on the second Sunday of June, Maryland politician
Bernie Fowler joins hands with scores of other people for a wade
into the Patuxent River to see what he can see through the murky water. The former state senator, Chesapeake Bay Commission member,
and chairman of the Patuxent River Commission is looking for his
feet, specially shod for the occasion in white sneakers. Back in the
1950s, when he was a waterman, Fowler could see his feet no matter
how deep he waded. In the decade that followed, the view grew
cloudier as the federal government financed sewage treatment plants
across the nation—some of them built upriver and pumping millions
of gallons into the Patuxent. By 1988, when Fowler initiated the annual wade-in, his feet disappeared after 8 inches. His goal is to be able
to see his feet in chest-high water.[1] By June of 1995, they were visible
after wading to a 40-inch depth—an improvement, but still about 20
inches short of the goal.

Through this "sneaker index," Fowler measures the progress of
efforts to restore the Bay. The Patuxent is an apt testing ground. The
river basin encompasses seven counties, where population more
than doubled and sewage treatment plant discharges increased 1,000
percent between 1960 and 1980. "Bernie Fowler Day," as the wade-in

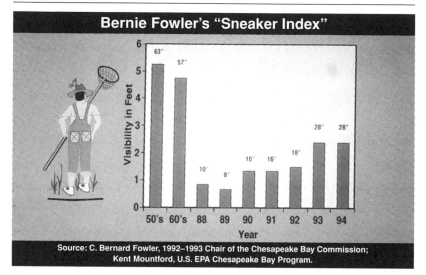

Bernie Fowler's "Sneaker Index"

Visibility in Feet / Year

63" 57" 10" 8" 16" 16" 18" 28" 28"

50's 60's 88 89 90 91 92 93 94

Source: C. Bernard Fowler, 1992–1993 Chair of the Chesapeake Bay Commission; Kent Mountford, U.S. EPA Chesapeake Bay Program.

As this chart illustrates, Bernie Fowler is getting closer each year to his goal of seeing his sneakers when he stands chest-high in the Patuxent River. Courtesy the Alliance for the Chesapeake Bay.

has come to be known, also takes the measure of Fowler's own environmental leadership.[2] In 1978, he engineered an agreement following a lawsuit filed by three Maryland counties challenging the state and the EPA over the water-quality plan for the Patuxent River. Victory by the counties in this lawsuit in 1981 produced a meeting of officials from the state and the seven counties, known today as the Patuxent River Charette, that set goals for water quality and aesthetic appearance and mandated reductions in daily discharges of nitrogen and phosphorus from sewage treatment plants into the river. Nitrogen removal systems subsequently added to the treatment plants were able to reduce discharges of that element by 18 percent, the best reduction of "point source" discharges of any Bay tributary. Discharges of phosphorus were cut 70 percent.

Leading the cleanup was the Patuxent River Commission, composed of representatives from the seven river watershed counties—Anne Arundel, Calvert, Charles, Howard, Montgomery, Prince George's, and St. Mary's—and state officials. Created by the legislature in 1980 to study the watershed and its pollution ills,[3] the commission concluded in 1983 that population growth and related changes in land use were the fundamental causes of pollution. Accordingly, the commission called for regulation of lands with potentially high runoff into the river and its tributaries, maintenance of vegetative buffers along the banks, closer government management of storm water runoff, and protection of agricultural lands and forests. After approval of their recommendations by the legislature in 1984, the seven counties incorporated the proposed regulations into their local comprehensive land-use plans and pollution-control programs.

The successful collaboration of these local jurisdictions provided the conceptual basis for the 1984 Chesapeake Bay Critical Area Protection Act, which established a cooperative effort between Maryland and its local governments to control land development and runoff pollution into the Bay and its tributaries.[4] Virginia also enacted a statute in 1988 to protect the Bay's water quality from development and runoff pollution, but this statute, unlike Maryland's, does not explicitly seek to protect valuable Bay habitats. The statutes are part of larger efforts among Bay states to cooperate in Bay watershed management as they have realized that the Bay and its problems extend well beyond the powers and capacities of individual states and local governments.

The EPA had pushed for a concerted regional approach in its 1983 study of the Chesapeake Bay. Maryland and Virginia, however, had anticipated the recommendation by a few years. The two states created the Chesapeake Bay Commission in 1980 to study the Bay and recommend administrative and legislative actions. Five years later, Pennsylvania joined, providing a membership dominated by state legislators.[5] The commission has since urged specific actions dealing with land use, fisheries management, pollution control, storm water management, sewage treatment, and protection of the Bay from nonindigenous species. A number of the commission's ideas have been adopted by the Bay Program.[6]

With this interstate environmental protocol already in place when the EPA released its study in the summer of 1983 to much fanfare in the news media, officials—from the governors on down—of Maryland and Virginia began consulting each other on how to respond. The governor of Pennsylvania and the mayor of Washington, D.C., were also invited into the conversation. The plan devised was to sponsor a governors' conference on the Chesapeake Bay in December 1983 to develop a joint federal-state Bay program.

THE 1983 GOVERNORS' CONFERENCE ON THE CHESAPEAKE BAY

The Chesapeake Bay conference in 1983 assembled business leaders, citizen activists, and government officials from Maryland, Virginia, Pennsylvania, the District of Columbia, and the EPA to come up with concerted actions to protect and restore the Bay. The customary high-flown rhetoric was put to the test at the end of the three-day conference when each jurisdiction was asked to announce specific initiatives for more effective treatment of sewage and tighter curbing of runoff from farms and land development. In these pronouncements, known as the Bay Declarations, it quickly became apparent that, just as geography had allotted vastly different means of access to the Bay and its riches, the states' respective commitments to its recuperation would vary accordingly.

Maryland Governor Harry Hughes faced the issue head-on, referring to "Maryland's special duty to do more," which he presented in the form of proposed expenditures of $70 million in 1984 and similar amounts thereafter. "Maryland puts more stress on the bay than either Pennsylvania or Virginia. . . . and therefore must assume a greater share of the responsibility," he said.

Governor Charles Robb of Virginia put up $6 million worth of proposals, admitting it was "only a fraction of what we need to spend." The modest declaration was greeted with catcalls from environmental activists, but Virginia officials hailed it as a momentous step in the right direction. "We're not comparing dollars [with Maryland], we're so happy to be where we are," said Larry Minnock of Virginia's Council on the Environment. "A year ago no one in Virginia thought we'd get any commitment. This is a day of celebration."

Pennsylvania's offer of $2 million—half from federal funds—to control agricultural runoff won a similar indulgence. Few observers expected as much from an upstream state with no economic stake in the Bay and at that time in the grip of severe budget problems.[7]

EPA Administrator William Ruckelshaus would say only that he was "cautiously optimistic" about increased federal funding for Bay programs. Agency officials had earlier testified against a bill that would have appropriated $50 million for such programs.

Among other issues, the states' commitments entailed plans for sewage treatment plant controls, planting aquatic grasses, and curbing development. Additionally, they and the EPA agreed to set up a program, at a cost of $2.5 million a year, to monitor annually the quality of the water and the status of living organisms in the Bay.

But the most significant outcome of the conference was the agreement by the participating jurisdictions to form a partnership, the Chesapeake Bay Program, dedicated to three simple but daunting tasks: reducing pollution, restoring habitat, and achieving sustainable harvests of fisheries.[8] The Bay Program's comprehensive approach includes developing new oyster reefs and sanctuaries; establishing forest buffers along the Bay and its tributaries to control runoff of nutrients (phosphorus and nitrogen) and erosion; reducing toxic pollutant discharges and runoff of nutrients from farms; protecting wetlands and wildlife and fish habitat; and improving treatment at sewage treatment plants to reduce their discharges of nutrients. However the commitments of the various jurisdictions might vary over time, an organization was now in place to attract funding for concerted action. As *Washington Post* reporter Todd Shields

The practice of maintaining trees and other vegetation along the shores of the Bay and its tributaries prevents erosion and absorbs pollutants that otherwise would end up in the Bay. Courtesy the Maryland Chesapeake Bay Critical Area Commission.

observed, ". . . the Bay Program has been part research engine, part idea clearinghouse and part funding catalyst."[9]

Congress now appropriates millions of dollars annually to the Bay Program. The states then supplement that with their own money. In this way, the program "has leveraged hundreds of millions of dollars in state money" for Bay restoration projects, said Ann Swanson, executive director of the Chesapeake Bay Commission.[10] For fiscal year 1995, for example, Congress appropriated $21 million to the Bay Program, with $9.6 million of it going directly to the Bay states for runoff pollution control and planting of forest buffers. Another $6.7 million went to universities, local governments, and nonprofit groups carrying out various programs. The remaining $4.7 million financed federal activity such as monitoring water quality and conducting computer modeling.

To govern the program, the agreement set up the Chesapeake Executive Council, which now consists of the governors of Maryland, Virginia, and Pennsylvania, the mayor of the District of Columbia, the chair of the Chesapeake Bay Commission, and the regional administrator of the EPA. Under terms of the agreement, the parties were to "share the responsibility for management decisions and resources regarding the high priority issues of the Chesapeake Bay."

The 1983 agreement produced no immediate substantive programs or standards for Bay restoration; instead, the jurisdictions responded with fully funded measures of their own. As Virginia Governor Gerald Baliles, who succeeded Governor Robb, put it, the 1983 agreement

> . . . was understandably, quite general, a pledge to cooperate, to begin the journey to clean up the Bay. Of course there was initial skepticism. It was said the states and the District of Columbia could not effectively work together, that they would not put their money where their mouths were, that the federal government would not put its money on the line. But the jurisdictions began by acknowledging that the decades-old decline of the Bay would not be turned around overnight. Commitments were made to the long haul, to pooling resources, energies and talents into an unrelenting effort. Plans were developed, funds were committed. While quite general in nature, what's important is this: the signers of that first

agreement put in place a process—one which would advance in stages the determined pursuit of the Bay's restoration.[11]

MARYLAND'S EARLY RESPONSE

The session of the Maryland General Assembly immediately following the 1983 conference passed more than thirty bills addressing the Chesapeake Bay environment and endorsed interstate cooperation with the other Bay jurisdictions in carrying them out.[12] The centerpiece of the 1984 General Assembly's work on this issue was the Chesapeake Bay Critical Area Protection Act, a plan for more standard state and local control of development along the shores of the Bay and its tributaries. The "critical area" was a statutorily defined one-thousand-foot ribbon of land that was thought to influence most directly the health of the waters at its edge.

In its passage through the General Assembly, the critical area bill faced a formidable array of adversaries who depicted it as a radical measure that would amount to an unconstitutional taking of land rights without compensation.[13] Real estate interests objected to restrictions on development. State Comptroller Louis L. Goldstein, who owned hundreds of acres of shoreline in Calvert and Dorchester counties, argued the bill would burden the rights of property owners. Legislators from the Eastern Shore counties—all of them are rural—most heavily affected by the bill were led in opposition by Senator Frederick C. Malkus of Dorchester County, whose rallying cry was: "Anytime you take a person's land and say what he can and can't do with it, it's wrong." They argued that, by choking off desired development on the Eastern Shore, the bill would unfairly single out that region to pay the price for pollution created upstream by industrial areas such as Baltimore.[14] The odds of passage were long at first.[15]

As originally proposed by Governor Hughes, the bill would have required local governments in Maryland to restrict development on land within 1,320 feet of the Bay and within 660 feet of minor tributaries. In a concession to opponents, the governor amended the critical area boundary to 1,000 feet. His strategy was to portray a vote against the bill as a vote against the Bay, a dangerous position for many legislators.[16] Among other arguments marshaled for enlarging the state role in local planning was the perception that many local planning and zoning decisions take place within a closed company of insiders, where powerful county officials effectively approve proposals before the public has its largely unheeded say. The critical area bill was supposed to inject environmental preservation principles into a process thought to be dominated by parochial interests.[17]

In addition to these arguments, the bill had a demographic advantage. Legislative reapportionment had diluted the strength of once-powerful sparsely populated rural senate districts, where much of the opposition was focused. Thus the senate was able to reject numerous proposals from rural legislators to weaken the bill and force them to end their filibuster.

"You're going to jam it down our throats," Malkus said in an emotional senate speech. "You want us to revert back to wilderness." He got the first part right. The bill passed the Senate forty-one to five on a roll call vote.[18]

The critical area governed by the legislation encompasses 641,613 acres, about 10 percent of the land area of Maryland. At the time of enactment, 42 percent of the undeveloped land in the critical area was forested, 37 percent agricultural. What made this statute unique at the time was its concept of a joint state-local approach, preserving local planning and zoning powers within the framework of state-directed standards.[19] Sixteen of Maryland's twenty-three counties and forty-four of its municipalities were required to draft a local program under the statute or let the newly created Chesapeake Bay Critical Area Commission[20] do it for them. Only the seven western-most counties of the state, which had no land within the defined critical area, were exempted. Under the statute, local governments remain primarily responsible for developing and enforcing their local versions of critical area regulation. But their programs and amendments enacted pursuant to the law are subject to review and approval by the commission. The state attorney general has authority under the act to file suit to enforce provisions of a local program applicable to any proposed development if the local government fails to do so.[21]

Baltimore Sun environmental writer Tom Horton welcomed the legislation but worried about the fate of the land immediately behind the protected strip of critical area. The law, he said, "is complex, and its regulations fill a small book, but the heart of the law says that in all the remaining undeveloped areas of the shore front, where we do not yet have severe environmental problems, we are going to dramatically restrict human activities to try and insure that we never do have problems to solve there. It is a very, very controversial piece of legislation. . . . However, many people are convinced it is a concept that must be expanded to the whole state, not just left to protect a thin fringe nearest the water, while progress as usual builds to the bursting point behind it."[22]

The state had stepped in to tighten what was perceived as laxity at the local level in regulating the Bay's vast waterfront. According to Senator Gerald Winegrad of Anne Arundel County, then the leading environmentalist in the senate, the legislature acted because local governments, with some exceptions, had "not exercised the planning and zoning powers so as to protect the loss of forest, farm land, and non-tidal wetlands directly around the bay and its tributaries." Without state action, he said, "the conversion from beneficial land uses in the 1,000 foot zone would undoubtedly continue."[23]

Each jurisdiction covered by the act drafted its own critical area plan, though many of the reluctant ones on the Eastern Shore simply followed the commission criteria word for word. That they cooperated at all is attributable in part to the skill and political savvy of Solomon Liss, whom Governor Hughes persuaded to resign as a judge of the Maryland Court of Special Appeals to serve as the commission's first chairman. Liss, an all-purpose politician who had served on the Baltimore circuit court and the city council before his appointment to the Court of Special Appeals, was often a calming influence as he presided over what might have been raucous public hearings on drafting the state criteria that local critical area plans would have to satisfy. At one such hearing in Leonardtown, in St. Mary's County, citizens urged—apparently facetiously—that if the criteria were truly to save the Bay, they must be applied throughout the watershed, from Virginia to New York. The affront to property rights was best summed up by a woman in attendance who told Liss, "I don't want to develop my land, but I don't want a commission to tell me not to develop my land."[24]

Liss, the urban pol, managed to contain the resentment seething at these hearings by presiding with the common sense and patience that reminded some observers of an old-time country doctor. Occasionally he offered soothing assurances that the commission he headed, that so offended notions of local control and individual property rights, "is not and never was intended to be a 'super-zoning board.' "[25]

How the Critical Areas Act Works

The act lists a minimum of eleven essential elements to any local critical area program, including a map of the affected area; limits to how much land will be covered by buildings, roads, and other impervious surfaces; incentives for cluster development as an alternative to suburban sprawl; shoreline buffer areas within which agriculture will be permitted if best management practices are employed; and minimum setbacks for structures and septic fields along the shore. The criteria seemed to empower local governments that wanted to get tough on waterfront development while they allowed less enthusiastic jurisdictions to interpret their way out of a strict regime.[26] Local governments were to apply these elements to three broad planning areas: (1) future land development, with the aim of protecting woodlands and farmlands so as to control erosion and absorb pollutants; (2) use and extraction of natural resources, primarily farming and timber cutting; and (3) protection of habitats.

When a private development project comes up for approval, the local authorities must base their decision on the critical area criteria.[27] Looking over their shoulder in the deliberations is the commission chairman, who has the right to intervene in any local administrative or judicial proceeding on the merits of a proposed project within a Critical Area.[28]

The critical area criteria govern future land development by requiring local governments to classify land within the critical area among three categories, according to the amount of development already existing and the availability of water and sewer lines in December 1985, the effective date of the criteria; and to limit new development within these three categories. About 5 percent of the critical area generally falls into the category of intensely developed areas (IDA);[29] 15 percent is limited development areas (LDA);[30] and the remaining 80 percent is resource conservation areas (RCA),[31] the most restrictive category, embracing about 320,000 acres of uplands and 200,000 acres of wetlands. New industrial and commercial development is generally prohibited in the RCA and new housing is limited to one dwelling per 20 acres.[32] But within the next category, LDA, most local jurisdictions permit new residential development at a density of four dwellings per acre, a density difference of 8,000 percent. As a result, classification as RCA or LDA was hotly contested in juris-

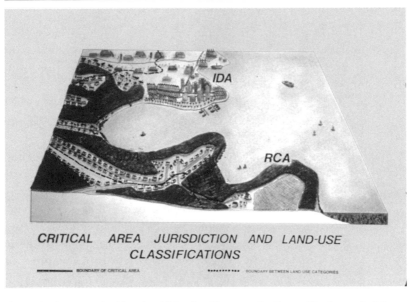

CRITICAL AREA JURISDICTION AND LAND-USE CLASSIFICATIONS

A particular parcel of land within the Chesapeake Bay critical area in Maryland is classified as an intensely developed area (IDA), a limited development area (LDA), or a resource conservation area (RCA). Courtesy the Maryland Chesapeake Bay Critical Area Commission.

dictions with extensive sewer lines.[33] Generally, new intense development must be located outside the critical area or in already intensely developed sections within it.[34]

Nevertheless, extensive development still occurs along the shoreline because of "grandfathering" provisions in the act that permit existing uses to continue inside the critical area, though they may not expand without a variance. These provisions also permit construction of single-family homes on lots that were legally recorded as of the date of commission approval of the local critical area program.[35] Moreover, the program sets forth growth allocation standards that allow a limited amount of intense development in the critical area, up to 5 percent of the acreage of a local jurisdiction's original resource conservation areas. About half of this allocation can involve conversion of land in the highly restrictive RCA category to LDA or IDA, and half conversion from LDA to IDA uses. The growth allocation formula exposes 18,495 acres in the sixteen critical area counties and Baltimore City to more intense development.

Now it was the environmentalists' turn for outrage. Saunders C. Hillyer, in 1988 the director of lands programs for the Chesapeake Bay Foundation, issued a report in that year lamenting how the politics of compromise had prevented the critical area legislation from imposing stiffer restrictions on development. The grandfathering provision and the growth allocation formula posed the prospect of an "unknown but potentially staggering" amount of development, which he blamed on the "give and take of the political arena." Though the precise number of grandfathered lots was unavailable, Hillyer estimated that tiny Talbot County alone had 1,600 such parcels.[36]

It fell to Critical Area Commission Chairman Solomon Liss to explain that politics was the art of what was possible to achieve among competing interests, not triumph for the ideals of any one of them. Perhaps the "scare tactics by developers," predicting an absolute halt to growth, had created a misconception about the act that was offensive to real estate interests, but pleasing to environmentalists. "Accommodate growth—that was the legislative instruction," Liss said, "but also to recognize that the more people that use the land near the shore of the Chesapeake, the more likely it is to be polluted." Between these diametrically opposed interests, "Who do you pick?" he said. "We have not fully pleased everybody, and that's a good sign."[37]

As with land use regulation, the criteria's governance of the use and extraction of natural resources looks slightly more accommodating on closer inspection. The permitted uses of natural resources include not only agriculture and mining but the construction of "water-dependent" facilities such as marinas and ports, public beaches, and fisheries operations.[38] In 1988, the Maryland General Assembly supplemented the criteria to prohibit drilling for oil or gas in the waters of the Bay or its tributaries.[39] A 1993 amendment further prohibited such activity in the critical area.[40] Maryland continues, however, to authorize permits for seismic operations in the Bay and its tributaries, or on the lands beneath those waters, to locate trapped oil or gas, providing the search poses no substantial risk of environmental damage.[41]

Under the Critical Area Commission's criteria for protecting habitats,[42] local governments are usually required to ban any development within a naturally vegetated buffer of at least 100 feet inland from the mean high waterline of tidal waters, tributaries, and tidal wetlands.[43] The buffers serve to remove nutrients (phosphorus and nitrogen) and other pollutants from stormwater runoff, to reduce erosion, protect wildlife habitat, and help maintain the natural environment of streams. Except for existing developed areas designated as buffer-exempt, new development is generally forbidden in the buffer, with the exception of private piers and a few other water-dependent uses.

A local critical area program is also required to protect the habitats of certain endangered and threatened bird species, riparian forests, and the spawning grounds of anadromous fish.

Impact of the Critical Area Criteria

The results of the critical area program are still flowing in, though patterns are emerging. The most noticeable is that it started a land rush. It started while the act was being debated in the General Assembly, as developers raced to file subdivision plans before the December 1, 1985, deadline for grandfathering lots.[44] By 1987, bayshore properties in communities such as Deale in Anne Arundel County were appreciating at 10 to 20 percent a year, compared with 5 to 7 percent before the critical area legislation. A state-financed Rutgers University study conducted for the Critical Area Commission confirmed such anecdotal evidence as reflective of trends throughout the Bay area. As a model for its analysis, the study established a theoretical "composite," ten-year-old, 2,000-square-foot house on a 2-acre waterfront lot. Based on an analysis of land sales from 1981 to 1986, the study concluded that buyers of such a house in 1986 were paying a "critical area premium" of anywhere from $648 in Dorchester County to $10,662 in Calvert.[45]

By 1990, speculators had bid prices beyond the reach of the middle class. The shores of the Bay were becoming a preserve of the wealthy. As one real estate broker explained it, the critical area law "has eliminated a fair number of people from the market. It used to be a family with an income of $65,000 to $70,000 could buy a decent [waterfront] home. Now, they're not even in the running. There are hardly any waterfront homes under $200,000. It seems the best-laid plans to 'Save the Bay' are merely accelerating gentrification. If the Chesapeake shores are being preserved, one can legitimately ask, 'For whom?' "[46]

The Critical Area Act had the opposite effect, however, on undeveloped lots that could not be grandfathered or allocated for growth.

A 100-acre tract in the most restrictive RCA zone that might once have yielded twenty or more buildable lots would now yield only five, making it far less attractive to developers and dropping its sale price precipitously.[47]

Despite the economic fallout, environmentalists welcomed the trend for its desired effect of protecting more land around the Bay. Saunders C. Hillyer of the Chesapeake Bay Foundation praised the critical area program in 1988 for installing "in embryonic form the fundamental elements of a successful Bay restoration program by providing for economic development, but only development that respects the watershed's ecological limitations, particularly the extraordinary sensitivity of the edge where land and water meet."[48] Tom Horton of the *Baltimore Sun*, however, noted that the law did not go far enough toward accomplishing what many environmentalists and scientists said was needed to protect the Bay: "namely, that some ultimate cap must be placed on the number of people who live near it."[49]

VIRGINIA'S 1988 CHESAPEAKE BAY PRESERVATION ACT

Virginia's Chesapeake Bay Preservation Act marked the first time the state had ever required local governments to regulate land use to protect the water quality of the Bay. Indeed, for some rural counties, the act forced the adoption of their first zoning ordinances ever. The statute, enacted in 1988, applied to twenty-nine counties, seventeen cities and forty-three towns in the Virginia tidewater. It largely followed the outline of Maryland's critical area act, but allowed for considerably more local discretion over the design and enforcement of the program.

The act established a state agency, the Chesapeake Bay Local Assistance Board, that promulgated criteria for local governments to follow in regulating land use around the Bay and its tributaries. But the jurisdictions were permitted to develop their local programs without having to gain approval of them from the state board, although in practice local Virginia jurisdictions submit their programs to the board for a determination that their program is consistent with the board's criteria. (The board may go to court to challenge a local program that it finds to be inconsistent with its criteria.) Nor are local jurisdictions bound to notify the board of proposed land developments, which means the board may have difficulty in ensuring compliance with the state criteria. This is part of the design, in keeping with Virginia's strong tradition of local control.

Generally, Virginia requires protection of only 100 feet inland from wetlands and tidal shores, compared with 1,000 feet in Maryland. Each local jurisdiction must classify lands according to a scheme similar to Maryland's IDA, LDA, and RCA, though the least restrictive category is optional under the Virginia statute.

The most controversial features of the board's criteria were the requirements for septic systems and vegetated buffers. The board relaxed its position on both. Proposed maintenance and construction language was deleted from the septic system criteria. Buffers of 100 feet are still generally required, but may be reduced to only 25 or 50 feet on agricultural lands. Developers were granted exemptions from the buffer criterion that in some cases permit homes and apartments to be built within the buffer area itself. Additionally, the board struck out provisions for protecting freshwater wetlands.

In contrast to Maryland's precise, formulaic limits on development, the board's criteria are vague in places, allowing for broad local interpretation. For example, development regulations provide that "no more land than is necessary . . . shall be disturbed," and that "vegetation shall be preserved to the maximum extent." Environmentalists naturally are critical. As Georgia Herbert of the Conservation Council of Virginia observed a year after enactment, "the regulations are full of weasel words. . . . It comes down to: 'Don't pollute if you don't want to.' "[50]

Impact of the Chesapeake Bay Preservation Act

The Virginia act is grudgingly being implemented at the local level. As of January 1995, seventy-eight of eighty-nine tidewater localities had adopted Phase I of the Bay preservation act programs, designating preservation areas and fashioning regulatory criteria for them. Some jurisdictions had got as far as Phase II, integrating water-quality measures into local comprehensive planning, and Phase III, incorporating those measures into zoning and other land management regulations as well. Because of this gradual compliance schedule and the wide discretion granted to local interpretation, no data are yet available to indicate whether the act is slowing development and reducing runoff pollution. As with the Maryland critical area legislation, however, the economic results registered immediately in real estate speculation. The cost of building a new home in the Virginia Chesapeake Bay preservation area rose anywhere from 0.5 percent to 3 percent due to the new regulations. Put another way, a $150,000 house went up in price by $1,400 to $4,500.

Other Virginia Enactments

The somewhat discretionary Chesapeake Bay Preservation Act is not the beginning and end of Virginia's response to the growing interstate and federal consensus for government intervention to safeguard the Bay. In 1989, Virginia enacted a prohibition[51] on oil and gas drilling in the Bay and its tributaries, or within 500 feet of the shoreline. Lands designated as resource protection areas under the Chesapeake Bay Preservation Act were similarly protected. That same year, the state amended its Coast Primary Sand Dunes and Beaches Act[52] to regulate development of beaches and dunes fronting the Bay. The amendment authorized five counties and three cities with sandy beaches fronting the Bay to adopt statutorily specified zoning ordinances generally outlawing use and alteration of such beaches and dunes without a permit from the local wetlands board or the state Marine Resources Commission. True to Virginia's preference for local control, the local jurisdictions were given the option, and were not required, to adopt the ordinances. As it turned out, all of them did.

THE 1987 CHESAPEAKE BAY AGREEMENT

The 1983 Chesapeake Bay conference was more important for its establishment of an intergovernmental structure and for laying out a common purpose than it was for any joint action. The 1987 Chesapeake Bay agreement carried the comity established at the 1983 conference to its next logical step—specific goals and programs. The signing ceremony, held at the Baltimore Convention Center with the pomp of red carpets, the Naval Academy Band, and a Coast Guard escort, was compared by the signers to a superpower summit. Behind the puffery and self-congratulation, however, was the realization that the task of restoring the Bay was so large as to humble proud, separate jurisdictions into working together. One of the signers of the 1987 Agreement, Virginia Governor Gerald L. Baliles, summed up its mood and purpose this way:

> The sheer size of the [Chesapeake Bay] watershed means the
> governments of hundreds of localities, several states, and the
> federal government must cooperate. Herding cats is easier. But
> without cooperation, dumping in one area might eliminate the
> gains produced by another locality's strict pollution controls.
> Furthermore, cooperation must be ensured over time. A clean
> Bay requires a commitment for the long haul. . . . So, any
> effective plan to restore the Bay had to ensure cooperation and
> outlast the terms of those who signed. The public nature of the
> 1987 agreement provided mutual assurances that
> commitments would be met. And the agreement offered
> specific and precise numbers and dates for many of the
> commitments. . . . Rather than create a single program, the
> agreement produced an ongoing process, a process designed to
> build upon itself.[53]

The agreement sets forth specific goals and objectives, and commitments to achieve them, in six areas: living resources; water quality; population growth and development; public information about the Bay, and education and participation in restoring it; public access to the Bay, its tributaries, and their shorelines; and governance. Perhaps the most heralded of the specific commitments toward achieving those goals was the pledge to reduce by at least 40 percent the amount of nitrogen and phosphorus entering the Bay by the year 2000. The resulting improvement in oxygen levels would improve the habitat of many of the Bay's living resources, such as crabs, oysters, striped bass, and submerged grasses. The 40 percent figure was derived from a computer model predicting a reduction of that size would eliminate the patches of total absence of oxygen in the Bay.

The Bay states attacked the nutrients problem on several fronts. The states' ban on the sale of laundry detergents containing phosphates brought a 30 to 50 percent reduction in phosphate discharges. Improved treatment methods and compliance with permit requirements by sewage treatment plants also have helped.

Runoff from farms was another obvious target in trying to stem the flow of nutrients into the Bay. In 1992, the Chesapeake Executive Council began promoting nutrient management plans and best management practices for agriculture. Farmers were urged to limit their use of fertilizers and try strip cropping, contour plowing, terracing of

Best management practices such as the strip cropping and grassed water-ways shown above help to reduce runoff of pollutants from farms in the Bay watershed. Courtesy the Maryland Chesapeake Bay Critical Area Commission.

land, construction of grassed waterways to trap sediment, and no-till and low-till practices to cut down on soil runoff. Pennsylvania put many of these suggestions into a statute enacted in 1993,[54] the first of its kind in the nation. It required farms with more than 2,000 pounds of livestock or poultry per acre per year to fashion a nutrient management plan and operate by it within five years. Under the statute, which applies to about 10 percent of the 21,500 Pennsylvania farms in the Bay basin, farmers must control use of chemical fertilizers and manure through proper storage and application. Terracing, planting winter cover crops, and other best management practices to control runoff are also part of the plans.

Maryland generally relies on its farmers to act voluntarily, encouraged by grants of nearly $3.5 million a year. The state offers to pay farmers 87.5 percent of the costs of installing best management practices that protect water quality. Farmers who take the state's money to build animal waste storage facilities must come up with a nutrient management plan in exchange. Additionally, all farms lying within the 1,000-foot critical area are required to have such a plan. As of June 1994, Maryland was nearly halfway toward achieving its goal of having such plans in place on 60 percent of the state's 2 million acres of cropland by the year 2000. The achievement may be on paper only, however, as Maryland does not check on whether the nutrient management plans are being implemented.

The Virginia approach follows the outline of Maryland's.[55] It also provides technical assistance to farmers.

1992 Amendments to the 1987 Agreement

By 1992, as the 1987 Chesapeake Bay agreement was making noticeable headway against nutrient pollution of the Bay, the parties to the agreement sought to include Bay tributaries in the program. The 1992 amendments recognized "a clear need to expand our program efforts in the tributaries, since most of the spawning grounds and essential habitat are in the tributaries." The parties also vowed to pursue cooperation with other states through which Bay tributaries flow: New York, West Virginia, and Delaware. The resulting interstate tributary strategy was based on computer models calculating the amount of nutrients (phosphorus and nitrogen) contributed by each tributary in 1985. From this data, the parties to the agreement and its amendments set goals for lowering nutrient contributions from each tributary so as to meet the overall reduction goal of 40 percent.[56] Major strategies were to reduce nutrient discharges into tributaries from sewage treatment plants and from runoff from developed land and farms (to a great extent through voluntary actions of farmers).

The amendments refined the nutrient reduction goal in the Bay to a permanent cap of 229.9 million pounds of nitrogen and 15.44 million pounds for phosphorus a year. That would maintain nutrient levels at 40 percent less than they had been in 1985, despite increased population and development in the Bay region. Achieving the goal would sometimes require cutting nutrient discharges by more than 40 percent to compensate for new sources arising from further development. Thus, Maryland sought to cut out 30 million pounds of nitrogen a year entering from its tributaries, even though the mandated reduction was only 22.7 million pounds, in order to accommodate the additional runoff expected from future growth and development. Virginia, Pennsylvania, and the District of Columbia followed a similar course.[57]

Key to attaining these goals is upgrading sewage treatment plants with the addition of biological nutrient removal technology, a process that uses bacteria to convert ammonia nitrogen in raw sewer water first to nitrate nitrogen and then to nitrogen gas. Maryland's strategy calls for upgrading about 40 sewage treatment plants with the technology, but the funding for such a costly project is as yet uncertain. Local government, business, and individuals are all likely to be tapped for the bill.[58]

The Bay states are well on their way to the goal of 40 percent less phosphorus discharges by the year 2000. By the end of 1994, Maryland had achieved 95 percent of the objective for phosphorus, Pennsylvania 49 percent, and Virginia 64 percent in the Potomac (figures were unavailable for the James and Rappahannock rivers). The District of Columbia has reduced only phosphorus discharges from its Blue Plains sewage treatment plant, and those only by 8 percent. Early in 1996, Virginia Attorney General James S. Gilmore III charged that the District had so neglected maintenance at the plant during the

Although discharges of phosphorus from the Blue Plains sewage treatment plant in the District of Columbia (shown) have decreased in recent years, Virginia Attorney General James S. Gilmore charged that improper management of the plant may result in discharges that could cause harm to the Potomac River and the Chesapeake Bay. Courtesy the Alliance for the Chesapeake Bay.

city's fiscal crisis that pollution controls were vulnerable to complete breakdown. The plant was "an environmental disaster waiting to happen," he said. After the U.S. government sued, the District signed a consent decree in April 1996 providing for changes in management and operation. But Gilmore opposed the decree, saying it still failed to force the plant to comply with its permit and did nothing to address the diversion of maintenance funds in the city's fiscal crisis. Over these objections, a federal judge approved the decree four months later.[59]

The other half of the Bay states' nutrients goal—stemming nitrogen discharges—has suffered slower going. By the end of 1994, Maryland had gone 58 percent of the distance to the nitrogen goal, Pennsylvania 21 percent, and Virginia 16 percent on the Potomac River. The District of Columbia had achieved 12 percent by the end of 1992. Nor have these modest reductions affected present levels in the Bay, which may now be 10 percent higher than 1984 levels. A more optimistic perspective on that figure takes into consideration that population and waste discharges have increased much faster. At this rate, however, the chief environmental engineer for the Chesapeake Bay Program, Clifford W. Randall, has concluded that the nitrogen goal will not be met by the year 2000, even if all planned agricultural best management practices and storm water controls are carried out.

Curbing the discharge of nitrogen has turned out to be more complicated and expensive than expected.[60] In Virginia, Maryland, and Pennsylvania strategies for farm waste and storm water control from development were estimated to cost $1.23 billion from 1995 to 2000. The states were expected to fall at least $272 million short. Moreover, the federal government no longer makes grants under the Clean Water Act for sewage treatment plant systems, although the District of Columbia secured a federal loan to finance most of a $12-million installation of new equipment at Blue Plains. As a result, only the District stands a chance of meeting its nitrogen deadline.

What progress there may be by the year 2000 may be overtaken by shifting calculations of how much nutrient reduction is needed to eliminate anoxia (absence of oxygen) conditions in the Bay. Recent computer modeling indicates that even if 90 percent of phosphorus and nitrogen were removed from the Bay, anoxic conditions would occasionally arise on account of spring freshets and tropical storms. With the data supporting its original goal undermined by these new figures, the Chesapeake Executive Council has requested research into new and better ways to cut down on phosphorus and nitrogen pollution.

FEDERAL AGENCIES CAN WORK TOGETHER, TOO

Shortly after the 1983 Chesapeake Bay conference, federal agencies acknowledged that they too needed a closer working relationship in addressing their disparate responsibilities regarding the Bay. In 1984, the Chesapeake Bay Program's Federal Agencies Committee was formed by the Environmental Protection Agency, the Soil Conservation Service, the Fish and Wildlife Service, the National Oceanic and Atmospheric Administration, the Geological Survey, the Army Corps of Engineers, and the Department of Defense.

Ten years later, twenty-nine federal agencies signed an Agreement of Federal Agencies on Ecosystem Management in the Chesapeake Bay, formalizing the increasing role of the federal government in the Chesapeake Bay Program. The agreement does not cover all the federal government programs supporting the goals of the Bay program but it does establish a framework for agency collaboration on pollution reduction, habitat restoration, operation of federal facilities, research, and ecological resource inventories. Some of the enunciated federal policies are unique to the Bay watershed, such as favoring forested buffers along streams, forming a tributary restoration strategy for federal lands in the District of Columbia, and upgrading sewage treatment plants at federal facilities to comply with nutrient reduction goals for Bay tributaries.[61]

GROWTH, SPRAWL, AND THE 2020 REPORT

For all the progress the Bay program can claim, new studies predicting yet more dire threats to the Bay continue to undermine hope for ultimate success. Foremost among these studies in pessimism is the so-called Year 2020 Panel report, written by experts commissioned under the 1987 Chesapeake Bay Agreement to examine the consequences of population growth and development into the next century. The experts projected that 59 percent more of the land in the Bay watershed would be developed by the year 2020 than was the case in 1980 and that 2.6 million new residents would be inhabiting it. The

When sprawl development occurs within the Bay watershed, runoff of harmful pollutants into the Bay and its tributaries increases because soil, trees, and vegetation are replaced by the impervious surfaces of driveways, streets, houses, and other structures. Courtesy the Maryland Chesapeake Bay Critical Area Commission.

A clustered development such as this one preserves more trees and vegetation than conventional sprawl development, thus resulting in less runoff of pollutants into the Bay and its tributaries. Courtesy the Maryland Chesapeake Bay Critical Area Commission.

217

report predicted that this pattern of unchecked growth "will cause serious damage to the Chesapeake Bay, unless state governments in the watershed adopt stringent measures."[62] Specifically what the panel had in mind were six "visions": (1) concentrating development in suitable areas; (2) protecting wetlands and other environmentally sensitive areas; (3) directing growth to existing settlements in rural areas (to take advantage of existing roads, schools, sewage systems, and other infrastructure); (4) making stewardship of the Bay a universal ethic; (5) practicing resource conservation, including a reduction in resource consumption, throughout the region; and (6) providing funding to pay for these plans.

The most controversial aspect of the report was the recommended means of enforcing these six visions: usurpation of local prerogatives by a statewide land use management scheme extending to all the land in the watershed. In 1991 a commission appointed by Maryland Governor William Donald Schaefer and headed by former Congressman Michael D. Barnes urged adoption of the Year 2020 Panel's six-point plan and statewide land management to enforce it. But farmers, real estate developers, and property rights advocates forged a coalition strong enough to defeat the proposals in the 1991 General Assembly.[63] A watered-down version of the legislation passed the following year. It requires county and municipal planning commissions in Maryland to include the Year 2020 Panel's six visions in their comprehensive plans and to adopt the necessary zoning and other ordinances to carry them out.

In Virginia, a specially appointed commission drafted legislation in response to the 2020 report in 1995, but the bills have not been enacted. In Pennsylvania, Governor William P. Casey denounced the proposals as "draconian."[64] Pennsylvania has yet to form a special panel to address the 2020 proposals, but a state legislative committee considering them has recommended giving the state a stronger role in managing growth and development.[65] As of 1996, however, the legislature had taken no action.

WILL THE JOB EVER BE FINISHED?

Assessing the work of twelve years of the Chesapeake Bay Program, the EPA likened it in a report to the condition of a hospital patient: "The doctor would report that the patient's vital signs, such as living resources, habitat and water quality, are stabilized and the patient is out of intensive care."

But it is also apparent that the Bay's recovery has proved to be a more complicated and prolonged task, and one costing more money, than anyone had anticipated.[66] As Chesapeake Bay Foundation President William C. Baker put it in the foundation's 1995 annual report: "A bright but unrealistic optimism in the early 1980s fueled hopes that the Bay could be saved relatively quickly. It now has been replaced by a determination to stay the course and to do whatever is necessary to get the job done."

Preserving that determination, and the cost of doing so, seems the next question to be answered. "The bottom line for all three [Bay] states is that they are going to need a lot more money," said Keith Gentzler, coordinator of Pennsylvania's Bay Program, "and right now I don't think any of us has a good idea where that will come from."[67]

Back at the Patuxent River, progress has slowed. In June 1996, Bernie Fowler's sneaker-shod feet disappeared after wading to a depth of 38 inches. That was about the same level as the previous year, and well short of Fowler's chest. "We still have a long way to go," he said, "but I still have high hopes we're going to make it one of these days."[68]

14

Perceptions of the Chesapeake Bay

[It is] a fair Bay compassed but for the mouth with fruitful and delightsome land. Heaven and earth never agreed better to frame a place for man's habitation.

—Captain John Smith, 1608

The Bay is the single most important, most valuable and significant aspect of the public features of Maryland. It is the essence and core of the Maryland we know and love. Its preservation and restoration are most, most important.

—Judge John North II, chairman, Chesapeake Bay
Critical Area Commission, 1990

The popular perception of the Chesapeake Bay has changed. This change is manifested in the journalistic accounts of the Bay. Originally it was viewed as a gift available to serve man's pleasure. Of late it has become an object of animistic concern, possessed of a natural life and an indwelling soul. Newspaper headlines ask, "Is the Bay Dead?" while the Chesapeake Bay Foundation exclaims, "Save the Bay!" The story of how the press, public opinion, and politics have interacted to change the Bay from an inexhaustible resource to a finite and endangered one affords a good summary of the recorded history of the Bay.

Early publicity treated the Bay as a bounty. On into the nineteenth century articles abounded that described the delights of the Chesapeake, the abundance of its resources, and the ruggedness of its watermen. However, in the course of this history bad news began to change the public's perception of the Bay. By the end of the nineteenth century germ theory had been accepted and with it the recognition of the Bay's "disease-engendering and pestilential conditions."[1] At the same time Bay cities were developing municipal water supplies and systems to carry off wastes from their growing residential and business districts. Bay tributaries adjacent to the tidewater towns became sinks into which these vast quantities of liquid were emptied.

Except for its effect on oyster culture, the use of the Bay as a disposal area seemed a happy and inexpensive solution to urban waste problems. At the turn of the twentieth century, when scientists indicted the oyster as a carrier of typhoid fever, the Bay oyster industry faced the marketing problem of how to quiet the concerns of the "pure food faddists." One response was the construction in Baltimore of a sewerage purification plant. Baltimore, the last major American city to install sewer lines, became the first to build a sewage treatment plant, on the Back River in 1914. The lobbyists who supported sewage treatment were more intent on cleansing the reputation of Maryland oysters than on cleansing the waters of the Bay.

The "oyster scare" publicity was not so easily dealt with, however. Epidemics in the mid-1920s, which were linked to raw oysters, led to water-quality and processing standards for oysters overseen by the U.S. Public Health Service. Maryland watermen continued to act as clean water lobbyists, with the result that the state was able to reduce oyster bed closures with a program of sewage treatment plant construction that led the nation. Virginia watermen were less powerful: Municipal wastes from tidewater towns poured untreated into receiving water. Virginia's legislature and courts refused to respond to the problem, and widespread oyster bed closures resulted.

One reason Maryland's oystermen proved more politically powerful than those of Virginia was that there were more of them. By 1900 Virginia had embraced a private oyster culture. Consequently, Virginia's oyster fishery was dominated by the relatively few leaseholders with larger, more efficient operations than in Maryland, where the oyster fishery remained open to all. The large number of Maryland oystermen composed a powerful lobby. Unfortunately, their large number was also destroying the natural oyster bars through overfishing. Despite editorials and feature articles lamenting the decline of a once-great industry (with titles such as "The Vanishing Oyster"[2] and "Oyster Stew"[3]), the legislature refused to limit access. Again, the political clout of Maryland oystermen, in this context intent on preserving the status quo, proved greater than the persuasive power of the press.

While Maryland oyster businesses created an effective lobby for clean water, their interests—bacterial discharges which polluted oyster beds—were specialized and localized. Sport fishermen joined them in support of clean water, but their interests related primarily to pollution of small free-flowing streams rather than pollution in the Bay. Meanwhile, in port areas, oil was dumped from the bilges of vessels plying the Bay and mixed with the ubiquitous discharges from food-processing plants. Industrial discharges in the Curtis Bay area of Baltimore and in portions of the York River attracted some attention, but, in the absence of a fish kill, little effective action resulted. During the first half of the twentieth century no group argued rigorously the case for good water quality throughout the Bay.

In the post–World War II years, scientific and technical work advanced understanding of water-quality issues. By the early 1960s new pollution sources had begun to share the long-standing attention given to municipal and industrial wastes. The effects of sediments and nutrients directed attention to activities throughout the Bay watershed, while problems associated with the disposal of dredged material became a matter of intense interest, particularly in Maryland. By the mid-1960s, waste heat from electric generating plants had also became a prime concern in Maryland. At about the same time, controversies arose in Maryland and Virginia over a number of engineering projects that had the potential to alter the circulation and salinity patterns of biologically significant areas in the Bay. As well, throughout the 1960s, the Washington metropolitan area wrestled with the problem of how to dispose of the wastes of a rapidly growing metropolitan area, while at the same time restoring the quality of the badly polluted upper Potomac estuary so that it could be used for recreation.

Biologists and physical scientists then became active in Bay affairs. They were called upon to study fish kills and declines in species and to project the likely effects brought on by public works projects, new industry, or waste disposal plants. Scientists also began to play key roles on boards and commissions that made policy recommendations to politicians and the public. They sought and received public funds for research and for monitoring public programs. Finally, they made economic, political, administrative, and legal decisions.

Scientists became advocates of clean water and constituted the steady force in the clean water lobby, although in particular cases they were joined by special stakeholders. These intervenors—opposed, for instance, to a transmission right-of-way or a diked spoil disposal site—had the good public relations sense to generalize their concerns, to argue that a new power plant would result in thermal pollution or that a dike failure could result in widespread dispersal of heavy metals.

The results of their efforts soon became apparent. In the early 1960s, a new kind of popular-press article came to be written which raised questions about Bay health and discussed in varying degrees of detail the problems facing Bay management. A sample of titles reflects their tone and content: "Does Pollution Threaten the Bay?"[4] "The Bay: An Abused Treasure,"[5] "Great Ugly Changes Mar Chesapeake,"[6] and "Bay Dangerously Polluted Despite Official Denials."[7]

These articles were remarkably similar: Almost without exception, they spoke of the once-boundless treasures of the Bay as reported by early settlers. They detailed the abundance of the Bay's resources, particularly the great oyster harvests of the late nineteenth century. They recounted the declines of commercial and sport species of fish and shellfish; they itemized specific threats to the Bay in the form of projects such as the nuclear power plant at Calvert Cliffs or the Chesapeake and Delaware Canal, as well as more generalized threats such as increasing urbanization and industrialization. They quoted Bay scientists and administrators, and they invariably made reference to the "health" of the Bay, often comparing it in some way with reputedly "dead" bodies of water such as Lake Erie, Lake Michigan, or Delaware Bay. They ended with a pessimistic prognosis for the Bay.

As indicated, these articles had shifted attention to a concern for the overall health of the Bay. They reflected the fact that the scientific community had become engrossed in ecology and its emphasis on the relationship between organisms and their environment. Scientists raised the public's consciousness that the Bay was a single system, and they were now joined by new allies—citizen environmentalists concerned with the overall productivity of the Chesapeake Bay ecosystem. In 1970 this perception found expression in the political arena when Maryland's Governor Marvin Mandel said: "We feel an almost sacred obligation to the Chesapeake Bay and its tributaries. . . . There is concern for the health of man and some parochial interests of those whose lives will be disturbed, but the predominant concern is for the Bay itself, the Bay as a living entity."[8]

The changing perception of the Bay was also reflected in the agencies that were now called upon to govern it. Sheriffs were the first governmental agents involved with the Bay. Legislation in both states empowered these local officials to regulate the harvesting of fish and to police the fishing of streams. With the "oyster wars" of the 1860s, Maryland and Virginia each attempted—through force of arms, on occasion—to keep the Bay's bounty for themselves. As fisheries took on greater economic significance in the late nineteenth century, both states formed boards to study, advise on, and eventually manage their fisheries.

These boards evolved into the Virginia Marine Resources Commission and the Maryland Tidewater Administration. This evolution has been relatively simple in Virginia, with only one simple name change in the last eighty years. In Maryland, constant legislative experimentation and organizational change produced at least ten different entities responsible for the management of tidewater fisheries resources. Despite these changes, attention to the needs of the sport and commercial fisheries appears to have remained.

In the latter part of the nineteenth century, both states formed boards of health which evolved into major state departments. These have retained responsibility for overseeing municipal sewage disposal and oyster sanitation. In Maryland, the state health agency has been a major factor in Bay management since the early 1900s. For forty years, it regulated both industrial and municipal waste treatment, with support from the various conservation agencies.

In 1939 Virginia created a regional sanitary authority for Hampton Roads, and by 1948 both states had statewide pollution control agencies. Virginia's Water Control Board had primary responsibility for both municipal and industrial waste control; in Maryland that responsibility was split between the health department and the Water Pollution Control Commission.

Maryland, in 1987, and Virginia, in 1993, established state departments of the environment. The Bay states have also created agencies devoted wholly to the protection of the Bay: at the state level, Maryland's Chesapeake Bay Critical Area Commission and Virginia's Chesapeake Bay Local Assistance Board; and at the interstate level, the Chesapeake Bay Commission. Of even more importance, the states and the federal government collaborate through the Chesapeake Bay Program.

The contribution of federal agencies began with the Army Corps of Engineers, which has had a continuing role in the development of

The Chesapeake Bay Program seeks to preserve beautiful scenic vistas of the Bay as well as to protect Bay water quality and living resources. Courtesy the Maryland Chesapeake Bay Critical Area Commission.

navigation facilities on the Bay. Since 1899 the federal government has regulated physical development on navigable waters and administered oil containment, nuisance weed removal, and debris removal programs. Most important to the water-quality story, the U.S. Public Health Service was an important source of early investigation of the Bay and later of supervising the shellfish sanitation programs of the states. After World War II, it administered the programs established by the water pollution control acts until this function was transferred to the Department of the Interior in 1965. Since 1970 the U.S. Environmental Protection Agency has been in charge of federal efforts to improve water quality and to protect and restore the Chesapeake Bay.

Still active are two research institutions on the Bay: the Virginia Institute of Marine Science and the Chesapeake Biological Laboratory. Johns Hopkins University's Chesapeake Bay Institute closed at the end of 1991 because of funding problems. All received a substantial amount of their financial support from state and federal governments, and the first two have been official state agencies for at least part of their histories. Their establishment (the first two around 1930, the latter in 1949) signaled the growing involvement of scientists in Bay governance.

In the first edition of this book, the authors, having surveyed the changing attitudes toward the Chesapeake Bay over nearly four hundred years, offered readers the following perceptions of their own:

- Much of the present discussion of governance of the Bay proceeds on the basis of an idealized model of how Bay management operates. The ideal or model first assumes preparation of a scientific baseline study that establishes the present condition of the Chesapeake Bay. Next comes the adoption of standards which choose the desired level of environmental quality for Bay waters. Then follows adoption of the comprehensive plan which selects the preferred uses of the Chesapeake Bay from among those consistent with the quality standards. Finally comes an implementation strategy whereby government puts the plan into effect through public works and private regulation. Our study shows that the reality is much different.
- The Chesapeake Bay may be the most studied estuary in the world, yet scientists have limited knowledge of its nature. But Bay scientists act as advocates and experts, bureaucrats and promoters. Although science so far has failed to provide a perfect baseline of information, scientists play a major role in shaping public perceptions of Bay problems.
- When making public choices affecting the quality of the Bay, government has responded more to public opinion and political pressure than to scientific analyses of its environmental and economic conditions.

Changes in approach to the Bay since 1983 require some modification of these three perceptions.

With respect to the first perception, the EPA's 1983 comprehensive study of the Bay created the scientific baseline envisioned by an idealized model. The Chesapeake Bay Agreements of 1983, 1987, and 1992 produced standards for comprehensive planning under the model, and the Chesapeake Executive Council provides the authority for implementation. Although the Chesapeake Bay Program will miss many of the deadlines for meeting goals of the Bay agreements, particularly for reducing nitrogen discharges by 40 percent, this approach to Bay management comes closer to the idealized model than anything that was in place when the first edition of this book was published.

Considering the second perception, it is probably no longer the case that scientists' knowledge of the nature of the Bay is limited, though it is still accurate to say that they lack a perfect baseline of information. The EPA's 1983 study, supplemented by subsequent monitoring and research, produced a substantial baseline of scientific knowledge. Even so, it remains a mystery whether the decline of the oyster is attributable to overharvesting or to the MSX and Dermo diseases. Nor is it yet clear whether the decline of the blue crab is owing to overharvesting or natural cycles in population. As for the role of scientists in shaping public perceptions of the Bay, it may be greater than ever: The news media frequently feature reports of data and conclusions from scientific monitoring and studies of the Bay.

Displayed on Maryland's "Treasure the Chesapeake" license plate, the great blue heron has become a symbol of public commitment to restoration and protection of the Chesapeake Bay. Courtesy the Maryland Chesapeake Bay Critical Area Commission.

Changes to the second perception affect the third. The wide dissemination of scientific findings through the news media frequently generates public pressure, evidence that the conclusions of science are not always at odds with the popular will. Rather, the issue has become how much governments can afford to spend for the solutions proposed by science, and what the effects of such solutions will be on existing economies and cultures dependent on the Bay. Where science is uncertain about an issue, such as what ails the oyster and the blue crab, political pressure—from watermen, environmentalists, and other Bay interest groups—determines the policy outcome.

Governments do not manage the Chesapeake Bay, as is often said. They lack the capacity. The Bay as a natural resource is too complex, knowledge is often too limited, and public choices are frequently too fickle. Susan Q. Stranahan, *Philadelphia Inquirer* journalist, recently observed that

> . . . the Chesapeake Bay represents the antithesis of a traditional environmental problem, and offers a view of environmental challenges that will arise in other venues into the next century. There is no single polluting industry that can be fined or shut down, no single resource, like water, that can be targeted for regulatory action. Nor will the problem be solved by a massive infusion of government dollars.[9]

This last statement, of course, represents apostasy to the authors of the many studies and reports that have urged ever greater government intervention, at staggering costs, on behalf of the Bay. Stranahan cites Caren Glotfelty, deputy secretary for water-quality programs within the Pennsylvania Department of Environmental Resources, in agreement that the problem goes beyond government to require a change in social attitude, from asking government to do more to people asking themselves how their conduct affects the Bay:

> The solution to such problems rests in "getting more people to claim ownership," to assume a stake in the solution . . . . "We need to get away from the idea that environmental protection comes down from Washington or Harrisburg. We've got to get people to understand that it's their personal choices that affect the environment—where they live, how they farm, how much

fertilizer they put on their lawns, small things like that." In short, it is no longer someone else's responsibility.[10]

Recognition that man has the capacity to alter the resources of the earth is not new. In 1863 George Marsh Perkins expressed this idea in his book *The Earth As Modified by Human Action.* Although this viewpoint did not come to occupy a central place in the history of science for another century, scientists from time to time have expressed concern for the Bay. In 1884 Dr. Charles W. Chancellor, health commissioner of Maryland, suggested, at least by implication, that the Bay could be ruined by the sewage from Baltimore.[11] In the 1930s the future of the Bay was discussed at the Chesapeake Bay Conference, and Dr. R. V. Truitt of the Chesapeake Biological Laboratory raised questions about the combined and cumulative effects of waste disposal.[12]

What is new is the widespread public and political concern for the future of the Bay. Former Virginia Governor Gerald L. Baliles, a signer of the 1987 Chesapeake Bay Agreement, believes, "Perhaps the most important progress made during the past decade has been the development of political and public understanding and support for the Bay." But Baliles also recognizes a paradox at work in the public's support of other initiatives that run at cross-purposes to public pieties about saving the Bay. The public desire for economic growth, he said, imposes "costs measured in lost rural lands, impure air and water, and damaged and declined natural resources . . . [and] in the enormous sums of public and private funds required to clean up the Bay, whether the economy is booming or in a state of decline."[13]

The ultimate question of whether man is ecologically endangering the Chesapeake Bay lies beyond the scope of this book and the knowledge of its authors, as does any specific recommendation for better government regulation of human activities that harm the Bay. Rather, the question raised by this book's survey of perceptions and management of the Bay is this: How much is society willing to sacrifice in public funds and changes in social and economic habits to achieve a more vibrant Bay? As efforts to safeguard the Bay's future exact a greater cost in terms of taxpayer dollars and the diminution of local authority and traditional Bay cultures, the challenge will be to fashion a public consensus that everyone has a stake in the Bay and that the cost of maintaining it can be spread equitably.

Notes

CHAPTER 1: FROM JAMESTOWN TO CONTRACT WITH AMERICA

1. C. Earle, "Early Virginia," 96.
2. Baliles, *Preserving the Chesapeake Bay*.
3. In 1964, the Maryland attorney general ruled that a statute when referring to "rivers, creeks or branches" did not apply to bays; this opinion was subsequently overruled by the Maryland Court of Appeals. *Board of Public Works v. Larmar Corp.*, 262 Maryland Reports 24, and 277 Atlantic Reporter 427 (1971).

CHAPTER 2: NOBLE ARM OF THE SEA

1. Justin Gillis, "Bay Meteor Theory Appears Rock Solid," *Washington Post*, 12 September 1995, A1, A8.
2. W. B. Cronin, *Volumetric, Areal and Tidal Statistics of the Chesapeake Bay Estuary and Its Tributaries*, Special Report 20 (Baltimore: Chesapeake Bay Institute, 1971). Despite the definitive nature of this report, many Bay authors have continued to use other sources for Bay dimensions, thereby perpetuating old errors.
3. U.S. Army Corps of Engineers, *Chesapeake Bay: Existing Conditions Report*, Appendix C: Processes and Resources C-VI-2. This report, which is the output of multiple authors spanning years of collection and compilation, actually contains several competing figures for various measurements for the Bay.
4. A. J. Lippson and R. L. Lippson, *The Condition of Chesapeake Bay and Assessment of Its Present State and Its Future*, paper presented to the Marine Environmental Quality Committee (Warsaw, Poland: International Council for the Exploration of the Sea, 1979), 1.
5. Particularly interesting discussion and analysis are contained in A. Y. Kuo et al., *Chesapeake Bay: A Study of Present and Future Water Quality and Its Ecological Effects* (Gloucester Point, Va.: Virginia Institute of Marine Science, 1975).
6. U.S. Army Corps of Engineers, *Chesapeake Bay: Future Conditions Report*, Summary, 19.
7. Ibid., 23.
8. Donald W. Pritchard, "Chemical and Physical Oceanography of the Bay," in State of Maryland, "Proceedings of the Governor's Conference," II-49.

9. L. Eugene Cronin, "The Biology of the Chesapeake Bay," in State of Maryland, op. cit., II-77 ff.

10. Kurt Blankenship, "Biodiversity and the Bay," *Bay Journal* 5, no. 6 (September 1995): 1, 6–7 (published by the Alliance for the Chesapeake Bay).

11. Bill Matuszeski, "Grand Canyon Experiment a Flood of Success," *Bay Journal* 6, no. 3 (May 1996): 17.

CHAPTER 3: MANY DEEP-WATER PORTS

1. See William J. Hargis, Jr., "Exploration and Research in Chesapeake Bay: Being a Brief History of the Development of Knowledge of the Bay of Santa Maria," Gloucester Point, Virginia, Institute of Marine Science Contribution, no. 837 (1977): 14–48.

2. Richard Hakluyt, quoted in Reps, *Tidewater Towns*, 24.

3. C. Earle, "Early Virginia," 103.

4. Percy, *Observations*, in Arber and Bradley, *John Smith*, lxii.

5. John Smith, quoted in Owens, *Baltimore on the Chesapeake*, 8–9.

6. The Virginia Council of the London Company, quoted in Reps, *Tidewater Towns*, 31.

7. Francis Makemie, "A Plain & Friendly Persuasive to the Inhabitants of Virginia and Maryland for Promoting Towns and Cohabitation," (1705), reprinted in *Virginia Magazine of History and Biography* 4 (1897): 261.

8. Padover, *Thomas Jefferson*, 326.

9. Population figures come from Simmons, *American Colonies*, 171.

10. Boorstin, *The Americans*, 106–7 (quoting John Clayton's 1688 letter to the Royal Society).

11. Starkey, *Land Where Our Fathers Died*, 11.

12. Cox, *Historic Alexandria, Virginia*, xiv.

13. The Reverend Hugh Jones, *Present State of Virginia*, London, 1724, quoted in Chesapeake Research Consortium, *Effects of Tropical Storm Agnes*, 66.

14. Boorstin, *The Americans*, 108.

15. Warren J. Coxe et al., *Guide to the Architecture of Washington, D.C.* (New York: Praeger, 1965), 3.

16. Coleman, *The First Frontier*, 270.

17. See Thomas J. Wertenbaker, "Intellectual Life around the Punch Bowl: Annapolis," in *The Golden Age of Colonial Culture*, 85–104.

18. Everstine, *The Compact of 1785*.

19. Boorstin, *The Americans*, 172; Billings, *Old Dominion*, 323.

20. Randall Beirne, interview by John Capper, December 1982.

21. Beirne, interview.

22. John H. B. Latrobe used this phrase about the following men he knew in his youth, just after the turn of the nineteenth century: Samuel Smith, Robert Gilmor, Robert Oliver, and other merchant-princes.

23. Beirne, interview.

CHAPTER 4: A WELL, A SINK, A PESTHOLE

1. John Smith, quoted in Burgess, *This Was Chesapeake Bay,* 4.
2. Maryland Laws, ch. 11 (1792).
3. Baltimore City Council, quoted in Scharf, *Chronicles,* 303.
4. Wertenbaker, *Norfolk,* 134.
5. Maryland Laws, ch. 79, sec. 13 (1808).
6. *Report of the Baltimore Water Commissioner* (Baltimore: James Lucas, 1853).
7. Maryland Laws, ch. 355 (1874).
8. Maryland Laws, ch. 6 (1886).
9. *Baltimore v. Warren Manufacturing Co.,* 59 Maryland Reports 96 (1882).
10. *Commonwealth v. Webb,* 6 Rand 27 Virginia Reports, 726 (1828).
11. *Miller v. Truehart,* 4 Leigh 31 Virginia Reports, 569 (1833).
12. 36 Southeastern Reporter, 373 (1900).
13. Ibid.
14. Charles Carroll the Barrister, in a letter to his London agent William Anderson, September 27, 1762, quoted in Michael Trostel, *Mount Clare* (Baltimore: Colonial Dames, 1982), 117.
15. Colonel Landon Carter, "The Diary of Colonel Landon Carter, of Sabine Hallan the Northern Neck of the Rappohannock River," *William and Mary College Quarterly* 13 (1904–05): 159.
16. Howard, *Disease in Baltimore,* 83.
17. Wolman, *Water, Health and Society,* 348–9.
18. Buckler, *Epidemic Cholera,* 30–31. See also Douglas Carroll, "History of the Baltimore City Hospitals, Chapter 5: The Almshouse at Calverton: The Beginning of Scientific Medicine (1840–1865)," *Maryland State Medical Journal* 15 (May 1966): 83.
19. Chesney, *Johns Hopkins,* 2:102–3.
20. See Armstrong, *Yellow Fever.*
21. Maryland Laws, ch. 93 (1801).
22. Maryland Laws, ch. 200 (1874).
23. Maryland Laws, ch. 12, sec. 1 (1886).
24. Van Bibber, "Drinking Waters in Maryland," 16–21.
25. Charles Varle, *A Complete View of Baltimore with a Statistical Sketch, etc.* (Baltimore: S. Young, 1833), 11.
26. Buckler, *Baltimore,* 10.
27. Buckler, *Epidemic Cholera,* 40.
28. Olson, *Baltimore,* 54, 92.
29. Maryland Laws, ch. 22 (1766).
30. Boorstin, *The Americans,* 238.
31. Thomas E. Bond, "Report of Baltimore Public Health Department, Baltimore, 1825," in *Baltimore City Health Department, the First Thirty-Five Annual Reports, 1815–1849* (Baltimore: Baltimore City Health Department, 1953).

32. Buckler, *Epidemic Cholera*, 20.

33. Abel Wolman, interview by John Capper, August 1980.

34. Quoted in Reps, *Tidewater Towns*, 216.

35. Leech, *Reveille in Washington*, 77.

36. Buckler, *The Basin and Federal Hill*, 49.

37. Hall, *Baltimore*, 424.

38. Campbell, *White and Black*, 257.

39. Buckler, *The Basin and Federal Hill*, 19.

40. Baltimore City Commissioners notation February 1768, document in Ms. 1242 in the Corner Collection, Maryland Historical Society, Baltimore.

41. Baltimore City Council resolution, June 1795.

42. Reps, *Tidewater Towns*, 126.

43. Kanarek, *Mid-Atlantic Engineers*, 117.

CHAPTER 5: TRADERS AND WATERMEN

1. Alsop, *Province of Maryland*, 450.

2. John Smith, quoted in Billings, *Old Dominion*, 215.

3. George Percy, *Account of the Voyage to Virginia and the Colony's First Days*, quoted in Billings, *Old Dominion*, 22–3.

4. William Gregory, "Journal from Fredericksburg," 226.

5. Power, *Chesapeake Bay in Legal Perspective*, 81–3.

6. Billings, *Old Dominion*, 321.

7. Maryland Laws, ch. 32 (1796).

8. Virginia Constitution, art. 11, sec. 3 (1971).

9. See Everstine, *The Compact of 1785*. Under the Compact of 1785 between Maryland and Virginia, full property rights along the shores were guaranteed to citizens of both states, including fishing rights. All laws to preserve fish had to be made with the consent of both states.

10. Maryland Laws, ch. 9 (1745).

11. Dorsey, "A Legal History," 4.

12. 6 Hen. 69, ch. 29, secs. 1–3 (1748).

13. Owners of mills were ordered to make slopes for the passage of fish by the following Virginia laws: Rivanna and Hedgeman rivers: 8 Hen. 361, sec. 81 (1769); Meherrin River, 8 Hen. 583, ch. 33 (1772); Rapidan River, 9 Hen. 579, ch. 39 (1778); 7 Hen. 321, ch. 32 (1759). Amended 7 Hen. 590, ch. 16 (for Appomattox River) (1762); 7 Hen. 423, ch. 20 (1761); 7 Hen. 321, ch. 32 (1759).

14. The Virginia legislature ordered Allan Howard's mill to be torn down because it entirely obstructed the passage of fish. 7 Hen. 423, ch. 20 (1761).

15. Patuxent: Maryland Laws, ch. 70 (1802); Monocacy: Maryland Laws, ch. 79 (1806); Susquehanna: Maryland Laws, ch. 91 (1813); Pocomoke: Maryland Laws, ch. 70 (1862).

16. Council Papers of Virginia, 1698, reprinted in *Virginia Magazine of History and Biography*, 21: 76 (January 1913).

17. Maryland Laws, ch. 95 (1810).

18. "Letters from John F. Mercer to Richard Sprigg," Mercer papers, April 19, 1797, in Pearson, *Fish and Fisheries*, 904–5.

19. Royall, *History, Life and Mariners*, 109–48 passim.

20. Ames, *Studies of the Virginia Eastern Shore*, 185.

21. State Planning Commission, *Conservation Problems in Maryland*, 3.

22. *Phipps v. Maryland*, 22 Maryland Reports 380 (1864).

23. *McCready v. Virginia*, 94 U.S. Reports 39 (1876).

24. Power, "More about Oysters," 202–10.

25. Maryland Laws, ch. 311 (1834).

26. Maryland Laws, art. 71, sec. 15 (1860); also ch. 184 (1867), on who may take what, where, when, and how.

27. Virginia Laws, 3 Hen. 46, 47 (1691).

28. Virginia Laws, ch. 36, sec. 13 (1748). Penalties were periodically imposed, including one for casting dead bodies into rivers, 3 Hen. 353, ch. 27 (1705).

29. Quoted in Dulaney, *The Architecture of Historic Richmond*, 2.

30. Quoted in Reps, *Tidewater Towns*, 222.

31. Virginia Laws, 2 Hen. 455, ch. 31 (1679).

32. Virginia Laws, 2 Hen. 484, ch. 15 (1680).

33. Virginia Laws, 4 Hen. 111, ch. 7 (1722).

34. Virginia Laws, 4 Hen. 177, ch. 7 (1726).

35. Virginia Laws, 4 Hen. 375, ch. 18 (1745).

36. *Harrison v. Sterett*, 4 Har. & McH. 540 (1774).

37. Maryland Laws, ch. 5 (1768).

38. Maryland Laws, ch. 24 (1783).

39. Maryland Laws, ch. 45 (1791).

40. Maryland Laws, ch. 45, sec. 7 (1791).

41. Virginia legislators responded with the following laws that provided for: clearing the Appomattox and Pamunkey rivers, 6 Hen. 394, ch. 40 (1752); clearing the James and the Chickahominy rivers and extending navigation, 8 Hen. 148, ch. 34 (1765); clearing the Great Falls of the James and extending navigation, 8 Hen. 148, ch. 34 (1765); extending navigation on the Potowmack [*sic*] from Fort Cumberland to the Tidewater, 8 Hen. 570, ch. 31 (1772); improving navigation, incorporation of the James River Company, 11 Hen. 450, ch. 19 (1784), and other improvements, 11 Hen. 341, ch. 25 (1783); opening and extending navigation of the Potomack [*sic*] and the Potomack Company incorporated, 11 Hen. 510, ch. 43 (1784); opening and extending navigation on the Appomattox, 12 Hen. 591, ch. 53 (1787), and on the Chickahominy, 12 Hen. 382, ch. 92 (1786).

42. Maryland Laws, ch. 35, sec. 3 (1800).

43. Maryland Laws, ch. 75 (1814).

44. Maryland Laws, ch. 154, sec. 18 (1817).

45. M. Gordon Wolman, "The Chesapeake Bay: Geology and Geography," in State of Maryland, "Proceedings of the Governor's Conference," II-7, -224.

46. Deric O'Bryan and Russell L. McAvoy, *Gunpowder Falls, Md.* (Washington, D.C.: Government Printing Office, 1966), 6.

47. Maryland Laws, ch. 27 (1753).

48. Maryland Laws, ch. 24 (1783).

49. Maryland Laws, ch. 58 (1872).

50. Olson, *Baltimore*, 55.

51. Lasson, "A History of Potomac River Conflicts," 27.

52. Wolman, "The Chesapeake Bay," in State of Maryland, "Proceedings of the Governor's Conference," vol. 2, 26.

53. *Garitee v. Baltimore*, 53 Maryland Reports 422 (1880).

54. *Baltimore and Ohio Railroad v. Chase*, 43 Maryland Reports 35 (1875).

55. *Norfolk City v. Cooke, Virginia*, 27 Gratt. 430 (1846).

56. Quoted in De Gast, *The Lighthouses of the Chesapeake*, 2.

CHAPTER 6: LAND OF PLEASANT LIVING

1. Seth and Seth, *Recollections*, 65–6.

2. Quoted in Robert and George Barrie, *Cruises* (Bryn Mawr, Pa.: Franklin Press, 1909), 51–2.

3. Brewington, *Chesapeake Bay*, 221.

4. Wertenbaker, *Norfolk*, 294–5.

5. Brewington, *Chesapeake Bay*, 223.

6. S. Earle, *The Chesapeake Country*, 265.

7. Seth and Seth, *Recollections*, 65–6.

8. Dr. John D. Godman, "American Natural History," quoted in John O'Ren, "Down the Spillway," *The Baltimore Sun*, 14 July 1955, 14.

9. Beadenkopf, "Baltimore Public Baths," 201, 205.

10. See Cooke, "Impact of Pollution."

11. Olmsted Brothers, *Development of Public Grounds*, 103–6.

12. Quoted in Reps, *Tidewater Towns*, 127–70.

13. Ibid., 75.

14. Ibid., 4.

15. Seth and Seth, *Recollections*, 64.

16. Brochure, Mount Vernon Ladies' Association of the Union, n.d.

17. Summerson, *Architecture in Britain*, 542.

18. Alan Gowans, *Images of American Living* (New York: Harper & Row, 1972), 133.

19. *Maryland Gazette*, August 20, 1754, quoted in Scharf, *The History of Maryland*, 2:13.

CHAPTER 7: AN IMMENSE PROTEIN FACTORY

1. There is no single study of the Bay oyster industry, although there is an abundance of source materials. For an outline of oyster legislation and production figures from 1810 to 1910, see Cumming, *Investigation of the Pollution of Tidal Waters*, 18.

2. See Brooks, *The Oyster*.

3. See, e.g., Governor Oden Bowie, *Message to the Maryland General Assembly* (Annapolis, 1870), 18–21; Governor Frank Brown, *Message to the General Assembly* (Annapolis, 1894), 16.

4. See Brooks, *Report of the Oyster Commission of 1884.*

5. For a detailed account of the industry see Hirschfeld, *Baltimore 1870–1900,* 42–62 passim.

6. State of Maryland, *Report of the Bureau of Labor and Statistics* (Annapolis, 1903), passim.

7. See Green, *Legislation of the State of Maryland,* 22–4.

8. See *Report on the Fisheries and Waterfowl of Maryland* (Annapolis, 1872), Document E, 37.

9. "Conservation Put under New Setup," *The Baltimore Sun,* 16 September 1941, 7.

10. See Frye, *The Men All Singing.*

11. Chesapeake Bay Authority, *Conference Report,* 121.

CHAPTER 8: SEWAGE AND SHELLFISH

1. *Manual of American Water Works, 1888* (New York: Engineers News, 1889), passim.

2. See Wolman and Geyer, *Report on Sanitary Sewers.*

3. For a particularly lively account of these issues in Baltimore, see Maryland State Board of Health, *Report,* 1884, 28–32, 140–78.

4. For an interesting summary of this process, see Martin V. Melosi, ed., *Pollution and Reform in American Cities, 1870–1930* (Austin: University of Texas Press, 1980).

5. Crooks, *Politics and Progress,* 132–54.

6. City of Baltimore, Sewerage Commission, *Annual Report,* 1897.

7. Maryland State Board of Health, *Biennial Report* 1886–87, 201.

8. City of Baltimore, Sewerage Commission, *Annual Report,* 1897, 55.

9. Speer, "Sanitary Engineering Aspects of Shellfish Pollution," 35.

10. City of Baltimore, Sewerage Commission, *Annual Report,* 1897, 54.

11. Ibid., 55.

12. Ibid., 80.

13. Crooks, *Politics and Progress,* 132–54.

14. "Mayor's Message to the City Council of Baltimore for the Year 1897" (Baltimore, 1898), 24. This document is housed in the Maryland Department of the Enoch Pratt Free Library.

15. Ibid., 25.

16. City of Baltimore, Sewerage Commission, *Annual Report,* 1899.

17. Ibid., 15–16.

18. Ibid., 17.

19. Maryland Laws, ch. 349 (1904), April 7, 1904.

20. City of Baltimore, Sewerage Commission, *Annual Report,* 1906, 17.

21. City of Baltimore, Sewerage Commission, *Report of the Board of Advisory Engineers,* 17.

22. Ibid., 43.

23. Commissioners of Fisheries of Virginia, *Annual Report*, Fiscal Year 1910–11, 11.

24. Commissioners of Fisheries of Virginia, *Annual Report*, Fiscal Year 1913–14, 14.

25. Commissioners of Fisheries of Virginia, *Annual Report*, Fiscal Year 1915–16, 12.

26. Commissioners of Fisheries of Virginia, *Annual Report*, Fiscal Year 1914–15, 12.

27. Commissioners of Fisheries of Virginia, *Annual Report*, Fiscal Year 1915–16, 12.

28. See *City of Hampton v. Watson*, 89 Southeastern Reporter, 81–3 (1916).

29. Ibid., 82.

30. Ibid.

31. *Darling v. City of Newport News*, 96 Southeastern Reporter, 307–15 (1918).

32. Ibid., 308.

33. Ibid., 309.

34. *Darling v. City of Newport News*, 249 U.S. Reports, 540–4 (1918).

35. Ibid., 542.

36. Ibid., 543.

37. Commissioners of Fisheries of Virginia, *Annual Report*, Fiscal Year 1922–23, 10.

38. Cumming, *Investigation of the Pollution of Tidal Waters.*

39. See Cumming, *Investigation of the Pollution and Sanitary Conditions of the Potomac Basin.*

40. Speer, "Sanitary Engineering Aspects of Shellfish Pollution," 24.

41. Office of Environmental Programs, Maryland Department of Health and Mental Hygiene, letter of October 29, 1925, in bound file titled "Notes and Information Concerning the Oyster Industry 1925–1928."

42. Maryland Department of Health, *Annual Report*, 1934, 7.

43. Chesapeake Bay Authority, *Conference Report*, 33.

44. Commissioners of Fisheries of Virginia, *Annual Report*, Fiscal Year 1928–1929, 11.

45. Commissioners of Fisheries of Virginia, *Annual Report*, Fiscal Year 1930, 6.

46. *Commonwealth v. City of Newport News*, 164 Southeastern Reporter, 690 (1932).

47. Ibid., 692.

48. Ibid., 689–700.

49. Ibid., 700.

50. Virginia General Assembly, *Pollution: Report of the Committee Appointed by the Governor.* Senate Document No. 6 (Richmond: Division of Purchase and Printing, January 1934), 3.

51. Ibid., 8.

52. Ibid., 9.

53. Ibid., 17–32.

54. The purpose of the act was ". . . the relief of the district from pollution and the consequent improvement of conditions affecting the public health and the natural oyster beds, rocks, and shoals."

55. Commissioners of Fisheries of Virginia. *Annual Report*, Fiscal Year 1938–1939, 9.

56. Ibid., 8.

CHAPTER 9: "DON'T LET THE FACTORIES IN"

1. In Virginia: e.g., Virginia Laws, 2 Hen. 484 ch. 15 (1680). In Maryland: e.g., Maryland Laws, ch. 79, sec. 13 (1808).

2. Maryland Commissioners of Fisheries, 6–7.

3. Cumming, *Investigation of the Pollution of Tidal Waters*, 43.

4. Maryland Laws, ch. 810 (1914), "The State Board of Health shall have general supervision and control over the waters of the State, insofar as their sanitary and physical condition affect the public health and comfort."

5. Maryland Laws, ch. 14 (1917).

6. Maryland Conservation Department, *Annual Report*, 1922, 24.

7. Ibid., 19.

8. Watson, "The Chesapeake Country's Life Revives," 1.

9. Maryland Department of Health, *Annual Report*, 1936, 182.

10. Maryland Conservation Department, *Annual Report*, 1936, 14–15.

11. "State Control of Pollution at New High," *The Baltimore Sun*, 3 October 1940, 28.

12. Maryland Department of Health, *Annual Report*, 1937, 195.

13. Ibid., 197.

14. Quoted in Maryland Water Pollution Control Commission, *Water Pollution*, 10.

15. Maryland Conservation Department, *Annual Report*, 1940, 59.

16. See articles in the "Water Pollution" vertical file in the Maryland Department at the Enoch Pratt Free Library in Baltimore.

17. Interview by John Capper of Dr. Cornelius W. Kruse, Department of Hygiene and Public Health, The Johns Hopkins University, Baltimore, October 16, 1980.

18. See F. C. Latrobe, "Would Unify Agencies: Maryland Outdoor Life Federation to Urge Constructive Program at Meeting Friday," *The Baltimore Sun*, 10 May 1936, sec. 2, 2; "Maryland Outdoor Life Federation Preparing to Push Fight for Conservation in State," *The Baltimore Sun*, 28 August 1938, sec. 1, 8; and "Outdoor Life Group Will Gather Tonight," *The Baltimore Sun*, 8 September 1938.

19. See, e.g., Acts of the Virginia Assembly, ch. 61 (1849–1850); ch. 147 (1852); ch. 85 (1874); ch. 270 (1884).

20. Levy, *Effect of Trade Wastes*, 21.

21. Maryland Conservation Department, *Annual Report*, 1922, 21.

22. Commissioners of Fisheries of Virginia, *Annual Report*, 1918–19, 10.

23. See discussions of oil pollution in Maryland Conservation Department, *Annual Report,* 1923, 27–31, and *Annual Report,* 1924.

24. Maryland Conservation Department, *Annual Report,* Fiscal Year 1930, 6.

25. Maryland Conservation Department, *Annual Report,* Fiscal Year 1938–39, 30.

26. Maryland Conservation Department, *Annual Report,* Fiscal Year 1940–41, 12.

27. Pleasants, *The Tidal James,* 108.

28. "Craney Island Spoil Disposal Area," briefing paper prepared by the U.S. Army Corps of Engineers, Norfolk District, June 1970.

29. Chesapeake Bay Authority, *Conference Report,* 1933.

30. Huntsman, "Oceanographic Research on Chesapeake Bay."

31. Ibid., 22.

32. Ibid., 14.

33. Ibid., 24.

CHAPTER 10: A BAY BUREAUCRACY

1. Commissioners of Fisheries of Virginia, *Annual Report,* Fiscal Year 1946–47, 9.

2. Commissioners of Fisheries of Virginia, *Annual Report,* Fiscal Year 1948–49, 30.

3. This account of the origins of the Water Control Board was obtained through an interview of A. H. Paessler, longtime executive secretary to the board, by John Capper, Richmond, December 1980.

4. Commissioners of Fisheries of Virginia, *Annual Report,* Fiscal Year 1948–49, 10.

5. Water Resources Policy Committee, *A Water Policy for the American People,* 187.

6. Paessler, interview (see note 3 above).

7. Commissioners of Fisheries of Virginia, *Annual Report,* Fiscal Year 1952–53, 10.

8. Commissioners of Fisheries of Virginia, *Annual Report,* Fiscal Year 1956–57, 10.

9. Commissioners of Fisheries of Virginia, *Annual Report,* Fiscal Year 1958–59, 11.

10. Commissioners of Fisheries of Virginia, *Annual Report,* Fiscal Year 1960–61, 12.

11. Commissioners of Fisheries of Virginia, *Annual Report,* Fiscal Year 1956–57, 49.

12. Ibid., 49–51.

13. Commissioners of Fisheries of Virginia, *Annual Report,* Fiscal Year 1960–61, 46.

14. Commissioners of Fisheries of Virginia, *Annual Report,* Fiscal Year 1958–59, 32.

15. Ibid., 51–2.

16. Maryland Board of Natural Resources, *Annual Report,* 1945, 76.

17. Ibid., 1947, 143.

18. See "State Pollution Study Is Slated," *The Baltimore Sun,* 27 November 1945, 14; "What Has Happened to the Water Pollution Committee?" *The Baltimore Sun,* 16 March 1946, 10; and "Agencies Back Pollution Bill," *The Baltimore Sun,* 3 October 1946, 12, for discussions of the development of the act.

19. Maryland Annotated Code Article 66C, secs. 34–45, 1951.

20. Maryland Water Pollution Control Commission, *Annual Report,* 1947, 9.

21. Maryland Water Pollution Control Commission, *Biennial Report,* 1952–53, 9.

22. Maryland Water Pollution Control Commission, *Biennial Report,* 1950–51, 25.

23. Dr. Joseph McLain, first chairman of the Water Pollution Control Commission, interview by John Capper, Chestertown, Maryland, September 1980.

24. Maryland Department of Health, *Annual Report,* 1949, 40–1.

25. "Plant Fights Pipeline Bar: DuPont Seeks to Extend Waste Dispoal Into Bay," *The Baltimore Sun,* 19 June 1957, 15.

26. Edgar L. Jones, "Maryland Pollution: Raw Sewage, Industrial Waste, Acid Mine Water Poured into the Chesapeake Bay and Streams," *The Baltimore Sun,* 6 March 1955, A3.

27. Odell Smith, "Pollution Plan Studied: State Aid Urged to Combat Condition," *The Baltimore Sun,* 12 March 1955, 15; and "Bay Pollution Study Group Is Appointed: Unit to Help Find Ways to Construct Disposal Plants," *The Baltimore Sun,* 29 November 1956, 42.

28. Maryland Department of Health, *Annual Report,* 1949, 49.

29. See Maryland Board of Natural Resources, *Annual Report,* 1947, 1959, and 1961.

30. "U.S. Proposes to Dump Mine Water into Bay," *The Baltimore Sun,* 1 March 1954, 26.

31. Wolman, Geyer, and Pratt, *A Clean Potomac,* 49–50. In 1957, a proposal was discussed at the American Water Works Association's annual meeting to clean up the Potomac by building a 30-mile, $55-million pipeline to carry effluent from the Blue Plains treatment plant to deep waters in the Chesapeake Bay. See "Anne Arundel Scorns Water Pollution Plan," *Washington Post,* 1 November 1957, B10.

CHAPTER 11: SAVE THE BAY: THE 1960s AND EARLY 1970s

1. Paessler, interview (see chapter 10, note 3).

2. References to these industrial waste symposiums appear in the annual reports of the Water Pollution Control Commission (after 1964 the Department of Water Resources) from 1961–68.

3. Commissioners of Fisheries of Virginia, *Annual Report*, Fiscal Year 1960–1961, 12.

4. Virginia Marine Resources Commission, *Annual Report*, Fiscal Year 1971, 24.

5. Virginia Marine Resources Commission, *Annual Report*, Fiscal Year 1972, 23.

6. State of Maryland, "Proceedings of the Governor's Conference," II-93.

7. See annual reports of the agency spanning this period.

8. See Pleasants, *The Tidal James*.

9. Nichols, "Salinity of the James River," 571 ff.

10. Virginia Marine Resources Commission, *Annual Report*, Fiscal Year 1972, 23.

11. Pleasants, *The Tidal James*, 109.

12. Code of Virginia (1950), sec. 28, 1–147.

13. Arnold, *Baltimore Engineers and the Chesapeake Bay*, 16.

14. David A. Maraniss, "Dredge or Die: The Port of Baltimore," *Washington Post*, 3 June 1979, A1.

15. Arnold, *Baltimore Engineers*.

16. Kanarek, *The Mid-Atlantic Engineers*, 117.

17. Maryland Board of Natural Resources, *Annual Report*, 1961, app. B, 151 ff.

18. Ibid., 153.

19. Maryland Board of Natural Resources, *Annual Report*, 1966, 8–9.

20. Letter of Joseph H. Manning, director of the Maryland Department of Chesapeake Bay Affairs, to Senator William James, April 16, 1971. Files of the Department of Natural Resources.

21. Maryland Department of Natural Resources Open File titled "Submerged Lands Commission" (1969).

22. Ibid.

23. This was extracted from various files of the Maryland Department of Natural Resources. The letters cited are: James B. Coulter to U.S. Army Corps of Engineers, Baltimore District, November 17, 1970, and Joseph H. Manning to Paul McKee, August 27, 1971.

24. Maryland Department of Chesapeake Bay Affairs, Wetlands Hearing File, Hart-Miller Islands Spoil Disposal Area (1971).

25. David Hoffman, "The Suffering Chesapeake: It's Not Dead Yet. But the World's Most Productive Estuary Could Be Killed by Politics and Pollution," *Washington Post Magazine*, 29 April 1979, 26.

26. *Hart and Miller Islands Area Environmental Group, Inc. v. Corps of Engineers of the U.S. Army*, 459 Federal Supplement Reporter, 279 (U.S. District Court, District of Maryland, 1978); *reversed*, 621 Federal Reporter, Second Series, 1281 (U.S. Court of Appeals, Fourth Circuit, 1980); *certiorari denied*, 449 U.S. Reports, 1003 (U.S. Supreme Court, 1980); *plaintiffs' motions for summary judgment denied and defendants' cross-motions for summary judgment granted,*

505 Federal Supplement Reporter, 732 (U.S. District Court, District of Maryland, 1980).

27. Maryland Annotated Code, Natural Resources Article, secs. 8-1601–8-1606.

28. Except for dredged spoil from local Baltimore County dredging projects, the Department of Natural Resources also has been prohibited since 1981 from approving any new contained area for the redeposit of dredged material taken from Baltimore harbor and its approach channels, within 5 miles of the Hart-Miller-Pleasure islands chain in Baltimore County. Any such area is prohibited from exceeding the 1,100-acre size of the Hart-Miller islands disposal facility.

29. Joe Nawrozki, "Dumping on Island Angers Eastside: State's Latest Decision for Hart-Miller Called Another Broken Vow," *The Baltimore Sun*, 16 June 1996, 1C.

30. David Hoffman, "The Suffering Chesapeake."

31. Angus Phillips, "Dredging Up the Facts on Poplar Island," *Washington Post*, 1 October 1995, D4.

32. Karl Blankenship, "Poplar Choice for Disposal, but Cost of Rebuilding Island from Sediment Raises Concerns," *Bay Journal* 6, no. 5 (July–August 1996).

33. Suzanne Wooton, "An Open Channel; Shipping Out: Baltimore's Position as a Leading East Coast Port Is Threatened by Difficulties Disposing of the Growing Amount of Dredge Produced by Maintaining Sea Lanes," *The Baltimore Sun*, 7 January 1996, 1D.

34. Suzanne Wooton, "Legislature Pressed for Dredge Disposal: Business, Labor Leaders Seek Reversal of Ban on Using Deep Trough," *The Baltimore Sun*, 9 February 1996, 1C.

35. See Natural Resources Institute, *Patuxent Thermal Studies*.

36. Maryland Department of Natural Resources, Office of the Secretary, Annapolis, "Calvert Cliffs Nuclear Power Plant" file, n.d.

37. D. W. Pritchard, "An Appraisal of the Probable Effects of the Morgantown Electric Power Plant on the Estuarine Environment of the Potomac River," undated mimeo, (letter of transmittal to Maryland State Senator Edward Hall, dated March 3, 1969), 1.

38. Ibid., 8.

39. Letter of L. Eugene Cronin to Donald W. Pritchard, March 6, 1969, 3, 4.

40. Ibid., 4.

41. Ibid., 6.

42. William M. Eaton, chairman, Nuclear Power Plants in Maryland, Report of the Governor's Task Force (Annapolis: December 1969).

43. Ibid., 51.

44. L. Eugene Cronin and Joseph A. Mihursky, press release, March 3, 1979. "Calvert Cliffs" file, Office of the Secretary, Maryland Department of Natural Resources, Annapolis.

45. Eaton, *Nuclear Power Plants in Maryland*, app. C, 2.

46. Ibid., app. D, 26.

47. Robert Bombay, "Calvert Nuclear Power Plant Ranked No. 1 Producer," *Baltimore News American*, 28 August 1976, 3A.

48. Baltimore Gas & Electric Company customer circular, "Energy News," May 1981.

49. See R. C. Rittenhouse, "Coal's Acid Test; New Attacks on Thermal Enhancement," *Power Engineering* 95, no. 9, (September 1991): 12.

50. Paessler, interview (see chapter 10, note 3).

51. Stranahan, *Susquehanna, River of Dreams*, 280.

52. Horton, *Bay Country*, 209.

53. Stranahan, *Susquehanna*, 288. The history of the creation of the commission is discussed on pp. 282–8.

54. Ibid., 288.

55. Donald W. Pritchard, interview by John Capper, Ocean City, Maryland, October 1980.

56. U.S. House Committee on Public Works, *The Chesapeake & Delaware Canal*, 91st Cong., 2d sess., 7, 8 April and 21 May 1970.

57. Ibid., 113–25.

58. Ibid., 273.

59. Ibid., 299 passim.

60. Letter, U.S. Army Corps of Engineers contracts office to Governor Marvin Mandel, October 8, 1970. Contained in the "Submerged Lands Commission" file, Department of Natural Resources Library, Annapolis.

61. See Arnold, *Baltimore Engineers and the Chesapeake Bay*, 46–50.

62. Maryland Board of Natural Resources, "Report of the Water Resources Committee," *Annual Report*, 1961, 144 ff.

63. Maryland Laws, ch. 82 (1964).

64. Federal Water Pollution Control Administration, *Chesapeake Bay: Susquehanna River Basin Project for Water Supply and Water Quality Management* (Washington, D.C.: Government Printing Office, 1965, revised June 1966).

65. William Colony, interview by John Capper, Crystal City, Virginia, September 1980.

66. U.S. Public Law No. 84-641, 70 Stat. 480 (July 2, 1956). See Arnold, *Baltimore Engineers and the Chesapeake Bay*, 31–4.

67. Angus Phillips, "Giant Chesapeake Model Is Closing Down," *Washington Post*, 11 March 1983, B1.

68. U.S. Army Corps of Engineers, *Chesapeake Bay: Existing Conditions Report* and *Future Conditions Report*.

69. William Eichbaum, "The Chesapeake Bay: Major Research Program Leads to Innovative Implementation," *Environmental Law Reporter* 14 (June 1984): 10,237, 10,238.

70. U.S. Department of the Interior, Fish and Wildlife Service, *National Estuary Study*, vol. 3 (Washington, D.C.: Government Printing Office, 1970), 66.

71. NASA, *Remote Sensing of the Chesapeake Bay*.

72. U.S. Public Law No. 94-116, 89 Stat. 588-89 (October 1, 1975). See Eichbaum, "The Chesapeake Bay," 10,239.

73. Horton, *Bay Country,* 218.

74. Wallace, McHarg, Roberts, and Todd, Inc., *Maryland Chesapeake Bay Study.*

75. See Pleasants, *The Tidal James,* and Hargis, "James River Basins."

76. See Wolman, Geyer, and Pratt, *A Clean Potomac River.*

77. Maryland Annotated Code, Natural Resources Article, title 8, subtitle 12, 1974.

78. Ibid., subtitle 11.

79. U.S. Public Law No. 92-500, sec. 208, 86 Stat. 839 (October 18, 1972), codified in *U.S. Code,* vol. 33, sec. 1288 (1973).

80. This is drawn from the personal experience of John Capper, who was hearing officer for approximately 150 Maryland wetlands cases in 1970–71.

81. Maryland Department of Natural Resources, Office of the Secretary, Steuart Petroleum file (1969).

82. Martha Angle, "Foes Grow on Waterways Industry: Citizens Fear Pollution," *Washington Evening Star,* 16 May 1969, H1.

83. Descriptions of the various dredging projects of the Harry Lundeberg School of Seamanship are contained in Maryland Department of Water Resources (now Water Resources Administration) waterway improvement files.

84. Maryland Annotated Code, Environment Article, title 16. Regulations under this Maryland statute (the Maryland Wetlands Act of 1970) are published in the Code of Maryland Regulations (COMAR) secs. 26.24.01 and 23.02.04.

85. Acts of the Virginia Assembly, ch. 711 (1972), codified, as amended, in Code of Virginia, secs. 28.2-1300–1320. This statute, the Virginia Wetlands Act, regulates development and other alteration of state and privately owned tidal wetlands located above the mean low-water (tide) mark. Virginia regulates development and other alteration of state-owned tidal wetlands located below the ordinary mean low-tide mark under a different statute: the Virginia Submerged Lands Act, Code of Virginia, secs. 28.2-1200–1213.

86. So called by Senator Gerald Winegrad. During his sixteen years of service in the Maryland House of Delegates and State Senate (from 1979 until 1995), Winegrad was regarded as the strongest advocate of protection for the environment in the Maryland General Assembly and was called the "environmental conscience" of the Maryland Senate by the *Washington Post.*

87. Timothy B. Wheeler, "Builder May Face Jail for Filling in Wetlands; Stiff Fines Could Make Company Go Bankrupt," *The Baltimore Sun,* 17 June 1996, 1B.

88. Ibid.

89. Dianne Dumanoski, "Maryland Wetlands Convict: Victim or Villain?" *Boston Globe,* 7 December 1992, 6.

90. Wheeler, "Builder May Face Jail."

91. Chesapeake Bay Foundation (CBF), "Wetlands Permitting Programs in the Chesapeake Bay Area" (Annapolis: CBF, 1994), 9. See Karl Blankenship, "CBF: Permit Process Slows Wetland Losses," *Bay Journal* 4, no. 9 (December 1994): 1, 10; Karl Blankenship, "Bay Wetland Losses Unabated in 1980s," *Bay Journal* 4, no. 2 (April 1994): 1, 7.

92. Chesapeake Bay Foundation, "Wetlands Permitting Programs," 9.

CHAPTER 12: ACTS OF GOD AND ACTS OF MAN

1. The Virginia Kepone episode was a subject addressed by the CBS program *60 Minutes* on December 14, 1975; Senate and House subcommittees held congressional hearings in early 1976 that focused on the Kepone episode in Virginia. See also Stan Crock, "Letter from Hopewell, Va.: The Poison Is Gone But the City Still Ails," *Business Week*, May 14, 1984, 34D.

2. Sandra Sugawara, "10 Years After Kepone Dumping, Problems Persist," *Washington Post*, 29 July 1985, C1. Professor William Goldfarb of Rutgers University has analyzed the Kepone pollution of the James River in "Kepone: A Case Study," *Environmental Law* 8 (1978): 645; and "Changes in the Clean Water Act Since Kepone: Would They Have Made a Difference?" *University of Richmond Law Review* 29 (1995): 603.

3. Phil McCombs, "Warm Weather Breathes Life into Bay," *Washington Post*, 7 March 1977, C1.

4. Mike Sager, "On the James Again: No Big Deal," *Washington Post*, 4 July 1981, C1.

5. See panel discussion, "Allied Chemical, the Kepone Incident, and the Settlements: Twenty Years Later," *University of Richmond Law Review* 29 (1995): 493, 508–9.

6. Morton Mintz and Daniel Klaidman, "Court Blasts Ex Parte Meetings with Allied Signal, Creative Settlement or Improper Deal," *Legal Times*, May 11, 1992, 1.

7. By 1992, the Virginia Environmental Endowment's assets had increased to $16 million. The endowment makes grants of approximately $800,000 per year for projects to improve the environment. In its eighteen years of existence, it has made grants totalling approximately $14 million.

8. Bill McAllister, "Allied's Fine Cut to $5 Million for Kepone Pollution," *Washington Post*, 1 February 1977, B1.

9. *Allied-Signal, Inc. v. Commissioner*, 63 T.C.M. (CCH) 2672 (1992). See *BNA Chemical Regulation Daily*, April 7, 1992. On February 23, 1995, the U.S. Court of Appeals for the Third Circuit affirmed this decision of the Tax Court. *Allied-Signal, Inc. v. Internal Revenue Service*, 54 Federal Reporter, Third Series, 767 (Third Circuit 1995).

Life Science Products, Inc., was convicted of 153 counts of federal criminal violations and criminally fined $3.8 million. This fine was not collected, however, because the company had become insolvent.

Some executives of Allied and the individual co-owners of Life Science Products also were convicted of federal crimes for their part in the Kepone pollution of the James River. In August 1976, one Allied executive pled guilty to a federal criminal charge of aiding and abetting illegal discharges of Kepone wastes into the James River. Another Allied executive pled guilty at that time to a federal criminal charge of making a false statement in a report to federal officials by omitting certain information about wastewater discharges. William Moore and Virgil Handtofte, Life Science Products' two co-owners and the company's president and vice president, were convicted of federal criminal charges, fined $25,000 each, and placed on probation for five years. Two Allied executives were acquitted, however, of federal criminal charges of conspiring to withhold, from EPA and the U.S. Army Corps of Engineers, data relating to Allied's discharges of Kepone into the James River.

The City of Hopewell pled nolo contendere (no contest) to ten of the 156 criminal counts in the federal grand jury indictments and was fined $10,000. The other 146 federal criminal charges against the city were dropped.

10. This settlement, reached in October 1977, resolved civil claims by the state and Hopewell for governmental costs incurred in responding to Kepone pollution of the James River, for damage to Hopewell's sewage treatment plant, and for state penalties. Of this settlement, $650,000 was paid to the City of Hopewell to repair the damage caused by Kepone to the city's sewage treatment plant. Another portion was allocated for Virginia's costs of monitoring the James River for Kepone. The allocation was exhausted by 1983.

11. *Pruitt v. Allied Chemical Corp.*, 523 Federal Supplement Reporter 975 (1981). The claims that were held to be legally recognized were based on negligence, strict product liability, nuisance, and admiralty law.

12. Goldfarb, "Changes in the Clean Water Act Since Kepone," 603, 618.

13. Chesapeake Research Consortium, *The Effects of Tropical Storm Agnes.*

14. Annie Dillard, *Pilgrim at Tinker Creek* (New York: HarperCollins Publishers Inc., 1974), 151.

15. Chesapeake Research Consortium, *Tropical Storm Agnes*, 15.

16. Karl Blankenship, "1994 Freshet Brings Wave of Ills to the Bay," *Bay Journal* 4, no. 6 (September 1994): 1. The large amounts of fresh water that ran into the Bay following the blizzard of 1996 and subsequent rains caused similar harm to the Bay. See Karl Blankenship, "Bay Water Flow 2nd Only to 1972," *Bay Journal* 6, no. 6 (September 1996): 1.

17. "Freshet Sequel: Low Flows to Bay Bring Some Surprises," *Bay Journal* 5, no. 5 (July–August 1995): 3.

18. William Eichbaum, "The Chesapeake Bay: Major Research Program Leads to Innovative Implementation," *Environmental Law Reporter* 14 (June 1984): 10,237, 10,243.

19. Ibid.

20. Karl Blankenship, "Detroit and Toronto Meet the Bay: Most Airborne Nitrogen Pollution Is Coming from Distant Sources," *Bay Journal* 5, no. 1 (March 1995): 1, 4–5.

21. Joe Coccaro, "EPA: Air Pollution Fouls Bay; Emissions Cited as Largest Single Source of Toxic Metals," *Virginian Pilot*, 26 April 1994, D1.

22. Karl Blankenship, "Toxics from the Sky: Researchers Examine Amounts of Airborne Toxics Reaching the Bay," *Bay Journal* 2, no. 9 (December 1992): 1.

23. Ibid.

24. Ibid.

25. Marjorie Sun, "The Chesapeake Bay's Difficult Comeback," *Science* 233 (August 15, 1986): 715.

26. The federal Clean Water Act essentially places water pollutants into four categories: conventional pollutants; toxic pollutants; nonconventional pollutants; and heat (thermal discharges). Under the Clean Water Act, conventional pollutants, which are the principal concern of primary and secondary treatment of wastes by publicly owned sewage treatment plants, include biological oxygen demand, suspended solids, fecal coliform, pH (a chemical measurement of acidity/alkalinity), oil, and grease. Nonconventional pollutants under the Clean Water Act include ammonia, chlorides, nitrates, iron, and color.

27. Karl Blankenship, "Defining the Bay's Toxic Problems," *Bay Journal* 4, no. 1 (March 1994): 1, 4–5.

29. See Karl Blankenship, "Impacts of Toxins on Bay's Wildlife Appear to Decline," *Bay Journal* 3, no. 3 (May 1993): 1.

29. See Bill Matuszeski, "To Clean Up the Bay, We Need To Look at the Air," *Bay Journal* 4, no. 1 (March 1994): 14; "Bay Benefits from Clean Air Act in Question as Law Is Challenged," *Bay Journal* 5, no. 1 (March 1995): 6–7.

30. D'Vera Cohn, "Maryland to Target Cars as Bay Pollution Source," *Washington Post*, 5 June 1991, A15.

31. Timothy B. Wheeler, "Car Smog Tests: The Price of Clean Air," *The Baltimore Sun*, 1 January 1995, 1C.

32. Ibid. After the 1997 regular session of the Maryland General Assembly, Governor Glendening vetoed a bill that would have prohibited dynamometer (treadmill) emission testing, leading the way for Maryland to begin such testing in October 1997.

33. "Glendening Seeks 65 MPH Limit on Md. Interstates, Rural Roads; Bill Also Would Raise Fines for Driving Too Fast, Ban Radar Detectors," *Washington Post*, 3 February 1995, C3.

34. Wheeler, "Car Smog Tests."

35. See Karl Blankenship, "Chewing Up the Landscape," *Bay Journal* 5, no. 9 (December 1995): 1, 6.

36. Lynne K. Varner, "Growing Green and Clean: U-Md. Studies Find lawn Chemicals Are Safe—If Used Properly," *Washington Post*, 26 July 1990, M3.

37. See D'Vera Cohn, "Bay Warming May Hinder Cleanup," *Washington Post*, 8 August 1991, C1.

38. "Chesapeake Bay Shoreline Erosion Study" [by the Corps of Engineers in 1991]. See Karl Blankenship, "Eroding Shores Reshape the Chesapeake," *Bay Journal* 1, no. 4 (June 1991): 1, 5.

39. See "Chesapeake Waterfowl Status and Trends," *Bay Journal* 4, no. 8 (November 1994): 14–5.

40. Karl Blankenship, "Bay Grasses Declined 10% in 1994; First Drop Since 1986," *Bay Journal* 5, no. 3 (May 1995): 1.

41. See Todd Shields and Angus Phillips, "Diseased Fish Caught in Potomac, Chesapeake Bay," *Washington Post*, 28 June 1997, B1; Todd Shields, "Md. Says Microbe May Be to Blame for Ailing Fish," *Washington Post*, 2 July 1997, B1.

42. Commercial oystering in the Bay, Maryland's regulation of oystering in the Bay, and Maryland's oyster replenishment program are discussed in Bill Gifford, "Shell Shock," *Washington Post Magazine*, 27 March 1994, 18.

43. See "Rockfish Rack Up Another Good Spawn in the Bay," *Bay Journal* 4, no. 7 (October 1994): 3.

44. For the history of shad restoration efforts in the Susquehanna River, see Stranahan, *Susquehanna, River of Dreams*, 243–77.

45. Todd Shields, "Blue Crabs in Sharp Decline in the Bay," *Washington Post*, 6 May 1995, A1, A11.

46. See "Blue Crab Decline May Date to 1972, Scientists Say," *Bay Journal* 5, no. 8 (November 1995): 6.

47. In 1994, Maryland banned the taking of "sponge" crabs (female crabs visibly bearing eggs) and also placed a limit on commercial crabbers of three hundred crabpots per individual license and nine hundred crab pots per boat. In addition, Maryland that same year imposed daily time limits on both recreational and commercial crabbing, a ceiling on commercial fishing licenses, and a limit of one bushel per day and two bushels per boat for recreational crabbers. Virginia did not similarly limit either the number of watermen or the number of crabpots that commercial watermen might use, but it has established crab sanctuaries, where crabbing is prohibited. In addition, Virginia in 1994 imposed a two-year waiting period for commercial crab licenses and restricted the winter crab fishery after December 1 (by placing limits on the size of dredges and the number of crabpots, and imposing a daily catch limit of 20 to 25 barrels). Recreational crabbing also was limited by Virginia, with five crabpots allowed to be used under a sport crabbing license (no license is required in Virginia for use of two crabpots and harvesting of one bushel of crabs per day). In addition, Virginia and Maryland require cull rings on crabpots to allow small crabs to escape from the pots. Both states adopted even stricter crab-harvesting regulations for 1995 and 1996, with Maryland shortening the harvesting season and prohibiting recreational and commercial crabbing altogether on specified days.

48. The ban would seek to protect mature female crabs while they undertake their annual fall migration to spawning areas at the mouth of the Bay.

This ban would principally affect commercial crabbers, who have been using more baited lines and more pots without getting larger harvests of crabs. This greater effort for equal or smaller crab harvests is considered by conservationists to be an indication that the Bay's crab population is reaching exhaustion.

49. Bill Gifford, "Fire and Water," *The Washington Post Magazine,* 28 July 1996, 17.

50. Timothy B. Wheeler, "Crabs Few, Despair Abundant; Smith Island," *The Baltimore Sun,* 12 May 1996, 1C.

51. Ibid.

52. Gifford, "Fire and Water." The Chesapeake Bay Foundation continues to maintain an active education center in Tylerton. While the climate has cooled somewhat since the difficult summer of 1996, there may always be a degree of tension. The arsonist has not yet been identified.

CHAPTER 13: MANAGING AN INTEGRATED ECOSYSTEM

1. "Bernie Fowler Day Shows Patuxent River Unchanged," *The Baltimore Sun,* 10 June 1996.

2. Angus Phillips, "Cleansing Patuxent Has a Wade to Go," *Washington Post,* 16 June 1992, D2.

3. Chapter 746, 1980 Laws of Maryland; presently codified, as amended, in Maryland Annotated Code, State Finance and Procurement Article, secs. 5-801–5-816.

4. Karl Blankenship, "Cleaning the Patuxent," *Bay Journal* 3, no. 5 (July–August 1993): 1, 11.

5. The Tri-State Agreement Creating the Chesapeake Bay Commission, as enacted by the legislature of each of the three states, is found in the Annotated Code of Maryland, Natural Resources Article, sec. 8-302; in the Code of Virginia, Title 62.1, ch. 5.2, secs. 62.1-69.5 through 62.1-69.20; and in the Laws of Pennsylvania, Act 25 of 1985, 32 P.S. sec. 820.11.

The commission, which has offices in Annapolis, Richmond, and Harrisburg, has twenty-one members, seven from each of these three states. Each state's members include two state senators, three state delegates or representatives, the state's governor or his or her designee, and a citizen representative.

6. See "Putting the Chesapeake Before the Legislature," *Bay Journal* 2, no. 2 (April 1992), 7.

7. Tom Horton, "Hughes Seeks Land-use Curbs along the Bay," *The Baltimore Sun,* 10 December 1983, A1.

8. Chesapeake Bay Foundation (CBF), *Save the Bay: 1995 in Review* (Annapolis: CBF, 1995).

9. Todd Shields, "For Bay Scientists, Budget Stalemate Means Months of Lost Opportunities," *Washington Post,* 3 February 1996, B5.

10. Ibid.

11. Baliles, *Preserving the Chesapeake Bay.*

12. See Solomon Liss and Lee R. Epstein, "The Chesapeake Bay Critical Area Commission Regulations: Process of Enactment and Effect on Private Property Interests," *University of Baltimore Law Review* 16 (1986): 54, 55–6.

13. Saunders C. Hillyer, "The Maryland Critical Area Program: Time to De-Mythologize and Move Forward," Chesapeake Bay Foundation memorandum, June 24, 1988, 3–4.

14. Gwen Ifill, "Md. Senate Backs Bay Development Curbs," *Washington Post,* 19 February 1986, C3.

15. Tom Horton, "Despite Noble Effort Bay Commission Failed Its Mandate," *The Baltimore Sun,* 18 August 1985, 3E.

16. Hillyer, "The Maryland Critical Area Program," 3.

17. Daniel Laskin, "Letter From Southern Maryland; Taking a Calm Look at a Touchy Issue," *Washington Post,* 13 December 1984, Md. 4.

18. Gwen Ifill, "Md. Senate Backs Bay Development Curbs," *Washington Post,* 19 February 1986, C3.

19. See Liss and Epstein, "Chesapeake Bay Critical Area Commission Regulations," 59.

20. The commission has twenty-seven voting members, each appointed by Maryland's governor. One serves as a full-time chairman. The voting membership must include eleven individuals who are elected or appointed officials of specified local counties and municipalities. Voting members of the commission also include eight individuals who are from specified counties and who represent diverse interests (such as commercial and recreational fishing, recreational boating, and environmental protection).

21. Maryland Annotated Code, Natural Resources Article, secs. 8-1815.1, 1815(b). The state's attorney general also may file suit in certain circumstances to enforce requirements of local approved project plans. Ibid, secs. 8-1815(d), (e).

22. Horton, *Bay Country,* 98.

23. Gerald Winegrad, "The Critical Areas Legislation: A Necessary Step to Restore the Chesapeake Bay," *University of Baltimore Law Forum,* Fall 1986, 3, 4, 5.

24. Daniel Laskin, "Letter From Southern Maryland; Taking a Calm Look at a Touchy Issue," *Washington Post,* 13 December 1984, Md. 4.

25. Ibid.

26. Horton, "Despite Noble Effort."

27. COMAR sec. 27.01.10.01. The commission is required to be notified of certain types of applications for project approval by a local jurisdiction. See COMAR sec. 27.03.01.

28. Maryland Annotated Code, Natural Resources Article, sec. 8-1812(a).

29. IDAs are defined by the Critical Area Commission as ". . . areas where residential, commercial, institutional, and/or industrial developed land uses predominate, and where relatively little natural habitat occurs." IDAs must have at least one of the following features: a housing density of four or more dwelling units per acre; a concentration of industrial, institutional, or commercial uses; or public sewer and water collection and distribu-

tion facilities currently serving the area and housing density exceeding three dwelling units per acre. In order to be classified as an IDA, these features also must be concentrated in an area of at least 20 adjacent acres or that entire upland portion of the critical area within the boundary of a municipality, whichever is less.

30. LDAs are defined as ". . . areas which are currently developed in low or moderate intensity uses" and which ". . . contain areas of natural plant and animal habitat, and the quality of runoff . . . has not been substantially altered or impaired." LDAs also must have at least one of the following four features: (1) housing density of between one dwelling unit per five acres and four dwelling units per acre; (2) lack of domination by agriculture, wetlands, forest, barren land, surface water, or open space; (3) public water or public sewer, or both; or (4) all the characteristics of an IDA except for the requirement of concentration in an area of at least 20 adjacent acres or a municipality's entire upland portion of the critical area.

31. Resource conservation areas (RCAs) are characterized by nature-dominated environments (wetlands, forests, or abandoned fields) and resource-utilization activities (agriculture, forestry, fishing, or aquaculture). RCAs must have at least one of the following characteristics: density of less than one dwelling unit per five acres, or dominant land use of agriculture, wetlands, forest, barren land, surface water, or open space.

32. This provision, one of the most controversial provisions of the commission's final criteria for local critical area programs, was adopted because of "evidence that linked low density land use to higher water quality and to preservation of terrestrial and aquatic habitats." Liss and Epstein, "Chesapeake Bay Critical Area Commission Regulations," 68. The commission adopted this one-house-per-20-acre density limitation after the commission's staff had recommended a density of one house per 50 acres.

This provision does not mean that each new residence in an RCA must be placed on a lot that is at least 20 acres in size. If a developer is subdividing and developing two or more lots in an RCA, he or she only has to achieve a density of one dwelling unit per 20 acres. For example, if a developer is subdividing a tract of 100 acres in the RCA, the one-residence-per-20-acres density requirement could be achieved by placing each of five houses on a 4-acre lot and leaving the remaining 80 acres undeveloped (and prohibited from ever being developed by a recorded, restrictive covenant). Furthermore, if a large tract in the RCA is located on the shoreline of the Bay or a tributary, a developer can place all the houses on the tract near the shoreline (leaving the required 100-foot vegetated buffer between the houses and the water [see pp. 206–7]).

33. Hillyer, "The Maryland Critical Area Program," 12.

34. Because of their potential for adversely affecting habitats or water quality, the following new activities or facilities are permitted within the critical area in IDAs only, and only after the activity or facility has demonstrated that there will be a net improvement in water quality to the adjacent body of water: (1) nonmaritime heavy industry; (2) transportation facilities

and utility transmission facilities (not including power plants), except where such location is necessary; (3) permanent sludge handling, storage, and disposal facilities, other than those associated with wastewater treatment facilities (however, agricultural or horticultural use of sludge may be permitted in the critical area, except in the 100-foot buffer).

In addition, because of their potential for adversely affecting habitat and water quality, new or expanded solid or hazardous waste collection or disposal facilities and new or expanded sanitary landfills may not be permitted in the critical area unless no environmentally acceptable alternative exists outside the critical area and the activity or facility is needed in order to correct an existing water-quality or wastewater management problem.

35. Furthermore, the "grandfathering" provisions in the commission's final criteria make the density requirements in the commission's final criteria inapplicable to: (1) any land on which, by December 1, 1985, development activity had progressed to the point of foundation footings or the installation of structural members; (2) any individual parcel of land that was recorded as of December 1, 1985, and that was not part of a subdivision; and (3) any lot that was subdivided into recorded, legally buildable lots prior to December 1, 1985. However, the design and construction of homes on such "grandfathered" lots must be in compliance with a local program's resource protection measures.

36. Robert Barnes, "Bay Foundation Scores Md. Critical Areas Law; Building Curbs Insufficient, Study Says," *Washington Post*, 24 June 1988, A11.

37. Ibid.

38. COMAR secs. 27.01.03–27.01.08.

39. Maryland Annotated Code, Natural Resources Article, sec. 6-104.4.

40. Under Maryland Annotated Code, Natural Resources Article sec. 6-104.2(b), a person may be able to obtain a permit from the state of Maryland authorizing the drilling of a production or exploratory well for oil or gas in an area outside the critical area using a method known as slant drilling, that drills through subterranean areas in the critical area, but an environmental impact study addressing the potential for any adverse environmental effects on the critical area from such drilling is required.

Virginia has enacted similar legislation that prohibits a person from drilling for oil or gas in the waters of the Chesapeake Bay or any of its tributaries and in specified areas of Tidewater Virginia along the shoreline of the Bay and its tributaries. Code of Virginia, sec. 62.1-195.1(A). See p. 210.

41. Maryland Annotated Code, Natural Resources Article, sec. 6-104.6. However, a person may not use explosives in seismic operations conducted on the waters of the Bay or its tributaries.

42. COMAR sec. 27.01.09.

43. Local jurisdictions are required to "expand the Buffer beyond 100 feet to include contiguous, sensitive areas, such as steep slopes, hydric soils, or highly erodible soils, whose development or disturbance may impact streams, wetlands, or other aquatic environments. In the case of contiguous slopes of 15 percent or greater, the buffer shall be expanded 4 feet for every 1

percent of slope, or to the top of the slope, whichever is greater in extent." COMAR sec. 27.01.09.01.C(7).

With the approval of the commission, local jurisdictions, as part of their local critical area program, may exempt portions of the critical area from the buffer requirements when the existing pattern of residential, industrial, commercial, or recreational development in the critical area would prevent a buffer in the area in question from fulfilling the buffer's functions. New development or redevelopment activities within buffer exemption areas are regulated by a final policy for buffer exemption areas adopted by the commission on May 5, 1993.

44. Niemeyer and Meyer, *Chesapeake Country,* 197.

45. Eugene L. Meyer, "Md. Law Drives Up Waterfront Prices; Rules Limit Critical Area Development," *Washington Post,* 7 August 1987, C1.

46. Niemeyer and Meyer, *Chesapeake Country,* 197–8.

47. See Meyer, "Md. Law Drives Up Waterfront Prices."

48. Hillyer, "The Maryland Critical Area Program," 34.

49. Horton, "Despite Noble Effort."

50. D'Vera Cohn, "Va. Tests the Waters on Bay Rules," *Washington Post,* 11 September 1989, A20.

51. Code of Virginia, sec. 62.1-195.1.

52. Code of Virginia, secs. 28.2-1400–1420.

53. Baliles, *Preserving the Chesapeake Bay.*

54. Purdon's Pennsylvania Statutes, title 3, secs. 1701–18.

55. In addition, large concentrated animal operations in Maryland and Virginia, which must have an NPDES Clean Water Act permit, are required to have a nutrient management plan.

56. See Karl Blankenship, "Nutrient Reduction Targets Are Set for Major Bay Rivers," *Bay Journal* 2, no. 8 (November 1992): 1, 6–7.

57. See "Achieving The Chesapeake Bay Nutrient Goals: A Synthesis of Tributary Strategies for the Bay's Ten Watersheds," EPA Chesapeake Bay Program Office, 1995; Karl Blankenship, "Keeping Pace with Nutrients," *Bay Journal* 4, no. 9 (December 1994): 1, 7; "Analysis Confirms Bay Nutrient Reduction Goal Is Attainable," *Bay Journal* 4, no. 8 (November 1994): 10.

58. Maryland's Blue Ribbon Panel on Financing Alternatives for Maryland's Tributary Strategies, established by Governor William Donald Schaefer in June 1994, sent a report, "Financing Alternatives for Maryland's Tributary Strategies," in early 1995 to new Maryland Governor Parris Glendening that identifies thirty-five techniques for funding the estimated $60 million per-year cost of Maryland's tributary strategies. See "Report Offers Options to Fund Md. Tributary Strategies," *Bay Journal* 5, no. 2 (April 1995): 3.

59. Toni Locy, "EPA-D.C. Blue Plains Plan Approved Despite Objections," *Washington Post,* 7 August 1996, D5.

60. Eric Lipton, "Cleanup Goal for Chesapeake Looks Doubtful," *Washington Post,* 11 December 1995, A1, A14.

61. See "The First Biennial Progress Report of the Agreement of Federal Agencies on Ecosystem Management in the Chesapeake Bay," Chesapeake

Bay Program's Federal Agencies Committee, April 1995; "Federal Agencies Commit to Restoring Bay Ecosystem," *Bay Journal* 4, no. 6 (September 1994): 6.

62. Phillip J. Tierney, "Maryland's 2020 Proposals: Strong Medicine for a Life Threatening Illness," *University of Baltimore Journal of Environmental Law* 1, no. 1 (1991): 24, 32.

63. Ibid., 34–7; Lawrence R. Liebesman and Karen M. Singer, "Maryland Growth and Chesapeake Bay Protection Act: The View from the Development Community," *University of Baltimore Journal of Environmental Law* 1, no. 1 (1991): 52–67.

64. Stranahan, *Susquehanna, River of Dreams,* 300.

65. See "Committee Calls for Greater State Role in Pa. Planning," *Bay Journal* 2, no. 6 (October 1992): 8; "Va. Growth Panel Plans Hearing on Draft Legislation," *Bay Journal* 2, no. 10 (January–February 1993): 5.

66. Lipton, "Cleanup Goal for Chesapeake Looks Doubtful."

67. Ibid.

68. "Bernie Fowler Day Shows Patuxent River Unchanged," *The Baltimore Sun,* 10 June 1996, Regional Digest.

CHAPTER 14: PERCEPTIONS OF THE CHESAPEAKE BAY

1. Buckler, *Epidemic Cholera,* 40.

2. Jack Yeaman Bryan, "The Vanishing Oyster," *The South Atlantic Quarterly* 48 (October 1949): 546.

3. Albert C. Lasher, "Oyster Stew: Maryland's Fishermen War Over $6.5 Million Yearly Bivalve Crop," *The Wall Street Journal,* 24 April 1953, 1. See Dr. R. V. Truitt, "There Aren't Enough Oysters," *The Baltimore Evening Sun,* 6 March 1935, 17; "Raw Bar Blues" (editorial), *The Baltimore Evening Sun,* 13 January 1939, 27; Frank Henry, "Bottom of the Oyster Barrel," *The Baltimore Sun,* 16 November 1947, A1; James Wharton, "Old Shellfish Story: Maryland's Oyster Shortages," *The Baltimore Sun,* 4 December 1958, 22.

4. John C. Schmidt, "Does Pollution Threaten the Bay? Constant Care Held Essential to Preserve State's Greatest Single Natural Resource," *The Baltimore Sun,* 28 July 1963, A1.

5. John Frye, "The Bay: An Abused Treasure," *The Baltimore Sun Magazine,* 21 April 1968, 6.

6. Tom Moore McBride, "Great, Ugly Changes Mar Chesapeake," *Baltimore News-American,* 17 August 1969, 1H.

7. Tom Cofield, "Bay Dangerously Polluted Despite Official Denials," *Baltimore News-American,* 20 April 1969, 1A.

8. Statement of Governor Marvin Mandel before the Subcommittee of Intergovernmental Relations of the Senate Committee on Government Operation, February 4, 1970.

9. Stranahan, *Susquehanna, River of Dreams,* 298.

10. Ibid.

11. Maryland Department of Health, *Report 1884,* quoting House and Senate Document I, 1884, ". . . in the course of time, Chesapeake Bay will become a distillation of all the filth of Baltimore City," 178.

12. See Maryland Conservation Department, *Annual Report,* 1940, 59; and Chesapeake Bay Authority, *Conference Report,* 1933.

13. Baliles, *Preserving the Chesapeake Bay.*

Selected Bibliography

Alsop, George A. *A Character of the Province of Maryland*. London, 1666. Vol. I. Edited by William Gowans. Baltimore: Maryland Historical Society Fund Publication, 1869.

Ames, Susie M. *Studies of the Virginia Eastern Shore in the Seventeenth Century*. New York: Russell and Russell, 1940.

Anderson, Alan D. *The Origin and Resolution of an Urban Crisis: Baltimore 1890–1930*. Baltimore: The Johns Hopkins University Press, 1977.

Armstrong, George D. *History of Yellow Fever in Norfolk*. Philadelphia: J. B. Lippincott & Co., 1856.

Arnold, Joseph L. *The Baltimore Engineers and the Chesapeake Bay, 1961–1987*. Baltimore: U.S. Army Corps of Engineers, Baltimore District, 1988.

Baliles, Gerald L. *Preserving the Chesapeake Bay*. Martinsville: Virginia Museum of Natural History Foundation, 1995.

Barbour, Philip. *The Jamestown Voyages under the First Charter, 1606–1609*. Cambridge, England: Cambridge University Press, 1969.

Beadenkopf, Anne. "The Baltimore Public Baths and Their Founder, the Rev. Thomas M. Beadenkopf," *Maryland Historical Magazine* 45 (1950), 201–14.

Beers, Roland, et al. *The Chesapeake Bay*. Springfield, Va.: National Technical Information Service, 1971.

Billings, Warren M., ed. *The Old Dominion in the Seventeenth Century*. Chapel Hill: University of North Carolina Press, 1975.

Blair, Carvel Hall, and Willits Dyer Ansel. *Chesapeake Bay: Notes and Sketches*. Cambridge, Md.: Tidewater Publishers, 1970.

Bodine, A. Aubrey. *Chesapeake Bay and Tidewater*. Baltimore: Bodine and Associates, 1954.

Boorstin, Daniel J. *The Americans: The Colonial Experience*. New York: Random House, 1958.

Brewington, Marion V. *Chesapeake Bay: A Pictorial Maritime History*. Cambridge, Md.: Cornell Maritime Press, 1956.

Brooks, William Keith. *The Development and Protection of the Oyster in Maine*. Baltimore: The Johns Hopkins University Press, 1884.

———. *The Oyster*. 2d ed. Baltimore: The Johns Hopkins University Press, 1905.

———. *Report of the Oyster Commission of 1884*. Baltimore: The Johns Hopkins University Press, 1884.

Browne, Gary. *Baltimore in the Nation: 1789–1861*. Chapel Hill: University of North Carolina Press, 1980.

Buckler, Thomas H. *Baltimore: Its Interest—Past, Present and Future*. Baltimore: Cushings & Bailey, 1878.

———. *The Basin and Federal Hill . . . Baltimore: Past Follies and Present Needs*. Baltimore, 1875. Monograph.

————. *A History of Epidemic Cholera*. Baltimore: J. Lucas, 1851.

Burgess, Robert H. *Chesapeake Circle*. Cambridge, Md.: Cornell Maritime Press, 1965.

————. *This Was Chesapeake Bay*. Centreville, Md.: Tidewater Publishers, 1963.

Byrd, William, II. *Prose Works: Narratives of a Colonial Virginian*. Edited by Louis B. Wright. Cambridge, Mass.: Harvard University Press, 1966.

Byron, Gilbert. *These Chesapeake Men*. North Montpelier, Vt.: Driftwind Press, 1942.

Campbell, Sir George, M.P. *White and Black: The Outcome of a Visit to the United States*. New York: R. Worthington, 1879.

Carlisle, George (Lord Morpeth). *Travels in America*. London, 1851.

Chesapeake Bay Authority. *Conference Report*. Baltimore, 1933.

Chesapeake Bay Critical Area Commission. *Critical Area and You: The Chesapeake's First Line of Defense*. Annapolis: Chesapeake Bay Critical Area Commission, undated.

————. *A Guide to the Chesapeake Bay Critical Area Criteria*. Annapolis: Chesapeake Bay Critical Area Commission, May 1986.

Chesapeake Research Consortium, Inc. *The Effects of Tropical Storm Agnes on the Chesapeake Bay Estuarine System*. Baltimore: The Johns Hopkins University Press, 1976.

Chesney, Alan M. *The Johns Hopkins Hospital and the Johns Hopkins University School of Medicine: A Chronicle*. 2 vols. Baltimore: The Johns Hopkins University Press, 1943.

Chowning, Larry S. *Chesapeake Legacy: Tools and Traditions*. Centreville, Md.: Tidewater Publishers, 1995.

City of Baltimore, Sewerage Commission. *Annual Report*. Baltimore, Fiscal Years 1897; 1899; and 1906.

————. *Report of the Board of Advisory Engineers and of the Chief Engineer on the Subject of Sewage Disposal*. Baltimore, 1906.

Coleman, R. V. *The First Frontier*. New York: Scribner's, 1948.

Commissioners of Fisheries of Virginia. *Annual Report*. Richmond, Fiscal Years 1910–11; 1914–15; 1915–16; 1918–19; 1922–23; 1928–29; 1930; 1938–39; 1946–47; 1948–49; 1952–53; and 1960–61.

Cooke, Daniel. "Impact of Pollution on the Water-Oriented Activities of Back River, Maryland." Master's thesis, University of Tennessee, Knoxville, 1968.

Cox, Ethelyn. *Historic Alexandria Virginia, Street by Street: A Survey of Existing Early Buildings*. Alexandria: Historic Alexandria Foundation, 1976.

Crooks, James B. *Politics and Progress: The Rise of Urban Progressivism in Baltimore 1895–1911*. Baton Rouge: Louisiana State University Press, 1968.

Cumming, Hugh S. *Investigation of the Pollution and Sanitary Conditions of the Potomac Basin*. Washington, D.C.: U.S. Public Health Service Hygienic Laboratory, Bulletin No. 104, 1916.

————. *Investigation of the Pollution of Tidal Waters of Maryland and Virginia, with Special Reference to Shellfish-Bearing Areas*. Washington, D.C.: U.S. Public Health Service Hygienic Laboratory, Bulletin No. 74, 1916.

Dabney, Virginius. *Richmond: The Story of a City*. New York: Doubleday, 1976.

De Gast, Robert. *The Bay*. Camden, Maine: International Marine Publishing Co., 1970.

———. *The Lighthouses of the Chesapeake*. Baltimore: The Johns Hopkins University Press, 1973.

———. *Oystermen of the Chesapeake*. Camden, Maine: International Marine Publishing Co., 1970.

Dorsey, Emerson L., Jr. "A Legal History of the Port of Baltimore." School of Law, University of Maryland Baltimore, 1978. Manuscript.

Dowdy, Clifford. *The Golden Age: A Climate for Greatness, Virginia 1732–1775*. Boston: Little, Brown, 1970.

Duke, Maurice, ed. *Chesapeake Bay Voices: Narratives from Four Centuries*. Richmond: Dietz Press, 1993.

Dulaney, Paul S. *The Architecture of Historic Richmond*. Charlottesville: University Press of Virginia, 1976.

Earle, Carville. "Environment, Disease, and Mortality in Early Virginia." In *The Chesapeake in the Seventeenth Century: Essays in Anglo-American Society*. New York: Norton, 1979.

Earle, Swepson. *The Chesapeake Country*. Baltimore: Thomsen-Ellis Co., 1929.

Eaton, William M. *Nuclear Power Plants in Maryland, Report of the Governor's Task Force*. Annapolis: State of Maryland, 1969.

Embrey, Alvin T. *Waters of the State*. Richmond: Old Dominion Press, 1931.

Evans, Ben. *Memories of Steamboating, Camp Meetings, Skipjacks and Islands of the Chesapeake*. Princess Anne, Md.: Marylander and Herald, Inc., 1977.

Everstine, Carl. *The Compact of 1785*. Annapolis: Legislative Council of Maryland, Research Report No. 26, 1946.

Footner, Hulbert. *Charles' Gift*. New York: Harper, 1939.

———. *Maryland Main and Eastern Shore*. New York: Appleton-Century, 1942.

———. *Rivers of the Eastern Shore*. Centreville, Md.: Tidewater Publishers, 1979.

Frye, John. *The Men All Singing: The Story of Menhaden Fishing*. Norfolk: Dunning, 1978.

Garitee, Jerome Randolph. *Private Enterprise and Public Spirit: Baltimore Privateering in the War of 1812*. Ann Arbor, Mich.: University Microfilms, 1973.

Green, Harry J. *A Study of the Legislation of the State of Maryland*. Baltimore: The Johns Hopkins University Press, 1930.

Greene, Suzanne Ellery. *Baltimore: An Illustrated History*. Woodland Hills, Calif.: Windsor Publications, 1980.

Gregory, William. "Journal from Fredericksburg, Virginia to Philadelphia." *William and Mary College Quarterly* 13 (1905).

Griffith, Thomas W. *Annals of Baltimore*. Baltimore: William Woody, 1824.

Guthein, Frederick. *Exploration and Research in Chesapeake Bay: Being a Brief History of the Development of Knowledge of the Bay of Santa Maria*. Gloucester Point, Va.: Virginia Institute of Marine Science, 1977.

———. *The Potomac*. New York: Rinehart & Co., 1968.

Hakluyt, Richard. *Discourse on Western Planning*. Edited by Charles Deane. Maine: Documentary History of Maine, n.d.

Hall, Clayton Coleman. *Baltimore: Its History and Its People*. New York: Lewis Historical Publishing Co., 1912.

Hargis, William J., Jr. "James River Basin: Great Natural Resource, or, Problems of Developing the James River." Gloucester Point ,Va.: Virginia Institute of Marine Science, 1963. Mimeographed.

Hays, Anne M., and Hazleton, Harriet R. *Chesapeake Kaleidoscope*. Centreville, Md.: Tidewater Publishers, 1975.

Hirschfeld, Charles. *Baltimore 1870–1900: Studies in Social History*. Baltimore: The Johns Hopkins University Press, 1941.

Horton, Tom. *Bay Country*. Baltimore: The Johns Hopkins University Press, 1987.

Horton, Tom, and William M. Eichbaum. *Turning the Tide: Saving the Chesapeake Bay*. Washington, D.C.: Island Press, Chesapeake Bay Foundation, 1991.

Howard, William Travis, Jr. *Public Health Administration and the Natural History of Disease in Baltimore, Md., 1979–1920*. Washington: The Carnegie Institute of Washington, 1924.

Huntsman, A. G. "Oceanographic Research on Chesapeake Bay." In the files of Maryland Department of Natural Resources. Annapolis, undated. Typescript.

Kanarek, Harold K. *The Mid-Atlantic Engineers: A History of the Baltimore District United States Army Corps of Engineers, 1774–1974*. Washington: Government Printing Office, 1977.

Land, Aubrey C. *Colonial Maryland*, Millwood, N.Y.: KTO Press, 1981.

Land, Aubrey C., Lois Green Carr, and Edward C. Papenfuse. *Law, Society, and Politics in Early Maryland*. Baltimore: The Johns Hopkins University Press, 1977.

Lang, Varley. *Follow the Water*. Winston-Salem, N.C.: John F. Blair, 1961.

Lasson, Kenneth. "A History of Potomac River Conflicts." In *Legal Rights in Potomac Waters*. Edited by Garrett Power. Bethesda, Md.: Interstate Commission on the Potomac River Basin General Publication 76-2, 1976.

Leech, Margaret. *Reveille in Washington*. New York: Harpers, 1941.

Levy, Ernest C. *Report to the Water Committee on the Investigation of the Effect of Trades Wastes on the Water of the James River at Richmond*. Richmond, 1905.

Lewis, Jack. *The Chesapeake Bay Scene*. Bridgeville, Del.: Jack Lewis, 1953.

———. *Potomac*. Dover, Del.: Woodwend Studios, n.d.

Liebmann, George. "The Chesapeake Bay Critical Area Act: The Evolution of a Statute." *The Daily Record*, Vol. 194, April 20, 1985, 1.

Lippson, Alice Jane, and Robert L. Lippson. *Life in the Chesapeake Bay*. Baltimore: The Johns Hopkins University Press, 1984.

Locke, Milo W. *Harbor of Baltimore: Locke's Plan*. Baltimore, 1875.

Maryland Board of Natural Resources. *Annual Report*. Annapolis, Fiscal Years 1945; 1947; 1959; and 1961.

Maryland Commissioners of Fisheries. *Annual Report*. Annapolis, Fiscal Years 1902 and 1903.

Maryland Conservation Department. *Annual Report*. Annapolis, Fiscal Years 1922; 1923; 1924; 1936; and 1940.

Maryland Department of Health. *Annual Report*. Annapolis, Fiscal Years 1934; 1936; 1937; and 1949.

Maryland Department of Health and Mental Hygiene. *Notes and Information Concerning the Oyster Industry 1925–1928.* Annapolis: Office of Environmental Programs, 1928.

Maryland State Board of Health. *Biennial Report.* Baltimore, 1886–87.

———. *Report.* Baltimore: House and Senate Document I, 1884.

Maryland Water Pollution Control Commission. *Annual Report.* Annapolis, Fiscal Years 1947; 1950; 1951; 1952; and 1953.

———. *Water Pollution: A Policy and Program for Control.* Annapolis, 1949.

Mayer, Brantz. *Baltimore Past and Present.* Baltimore: Richardson & Bennett, 1871.

McCloskey, William. "Hard Times Hit the Bay," *National Wildlife* 22 (April-May 1984): 7–14.

Mencken, H. L. *Happy Days.* New York: Alfred A. Knopf, 1940.

———. *Newspaper Days.* New York: Alfred A. Knopf, 1941.

Middleton, Arthur Pierce. *Tobacco Coast: A Maritime History of Chesapeake Bay in the Colonial Era.* Newport News: The Mariners' Museum, 1953.

Miers, Earl Schenck. *The Drowned River.* Newark, Del.: Curtis Paper Co., 1967.

Morgan, Edmund S. *Virginians at Home; Family Life in the Eighteenth Century.* Williamsburg, Va.: Colonial Williamsburg, 1952.

Morpeth, Lord. *See* Carlisle, George.

Mount Vernon. Mount Vernon, Va.: Mount Vernon Ladies' Association of the Union, 1978.

NASA (National Aeronautics and Space Administration). *Remote Sensing of the Chesapeake Bay: Conference Report.* Wallops Island, Va.: NASA, 1971.

"National Environmental Symposium on the Chesapeake Bay." *Maryland Law Review* 47, no. 2 (Winter 1988): 341–496.

Natural Resources Institute. *Patuxent Thermal Studies, Summary and Recommendations.* Special Report No. 1. College Park, Md.: Natural Resources Institute, 1969.

Nichols, M. M. "The Effect of Increasing Depth on the Salinity of the James River." Geological Society of America Contribution No. 382. Gloucester Point, Va.: Virginia Institute of Marine Science, 1972. Mimeographed.

Niemeyer, Lucian L. (photographs), and Eugene L. Meyer (text). *Chesapeake Country.* New York: Abbeville Publishers, 1990.

Olmsted Brothers, *Report Upon the Development of Public Grounds for Greater Baltimore.* Baltimore: Lord Baltimore Press, 1904.

Olson, Sherry. *Baltimore: The Building of An American City.* Baltimore: The Johns Hopkins University Press, 1980.

Owens, Hamilton. *Baltimore on the Chesapeake.* Garden City, N.Y.: Doubleday, 1941.

Padover, Saul K., ed. *Thomas Jefferson and the National Capital 1783–1818.* Washington, D.C.: Smithsonian Institution pamphlet, n.d.

Pearson, John C. *The Fish and Fisheries of Colonial North America.* Parts 4 and 5. Washington, D.C.: Fish and Wildlife Service, National Marine Fisheries Service, 1972.

Peden, William, ed. *Thomas Jefferson: Notes on the State of Virginia.* Chapel Hill: University of North Carolina, 1955.

Peffer, Randall S. *Watermen.* Baltimore: The Johns Hopkins University Press, 1979.

Percy, George. *Observations Gathered out of a Discourse of the Plantation of the Southern Colonie in Virginia 1606.* Reprinted in Edward Arber and A. G. Bradley, *Travels and Works of Captain John Smith,* vol. 1, lxii. London, 1625.

Pleasants, John B. *The Tidal James: A Review.* Special Report No. 18. Gloucester Point, Va.: Virginia Institute of Marine Science, 1971.

Power, Garrett. *Chesapeake Bay in Legal Perspective.* Washington, D.C.: Government Printing Office, 1970.

———. "More about Oysters Than You Wanted to Know." *Maryland Law Review* 30 (1970): 199.

Powers, Ann E. "Overview: Law: Protecting the Chesapeake Bay: Maryland's Critical Area Program," *Environment* 28, no. 4 (May 1986): 5.

Reps, John W. *Tidewater Towns: City Planning in Colonial Virginia and Maryland.* Chapel Hill: University of North Carolina Press, 1972.

Reynolds, Michael, et al., eds. *Maryland: A New Guide to the Old Line State.* Baltimore: The Johns Hopkins University Press, 1976.

Ridgway, Whitman H. *Community Leadership in Maryland 1790–1840: A Comparative Analysis of Power in Society.* Chapel Hill: University of North Carolina Press, 1979.

Risjord, Norman K. *Chesapeake Politics 1781–1800.* New York: Columbia University Press, 1978.

Royall, Anne. *Sketches of the History, Life and Manners in the United States.* New Haven, Conn., 1826.

Scharf, J. Thomas. *Chronicles of Baltimore.* Baltimore: Turnbull Brothers, 1874.

———. *History of Baltimore City and County.* Philadelphia: Louis H. Everts, 1881.

———. *The History of Maryland,* 3 vols. Baltimore: Turnbull Brothers, 1879.

Schubel, J. R. *The Life and Death of the Chesapeake Bay.* College Park, Md.: University of Maryland Sea Grant College, 1986.

Semmes, Raphael. *Captains and Mariners of Early Maryland.* Baltimore: The Johns Hopkins University Press, 1931.

———. *Tidewater Boy.* Indianapolis: Bobbs-Merrill, 1952.

Seth, Joseph B., and Mary W. Seth. *Recollections of a Long Life on the Eastern Shore.* Easton, Md.: The Press of *The Star-Democrat,* 1926.

Sherwood, Arthur W. *Understanding the Chesapeake, A Layman's Guide.* Centreville, Md.: Tidewater Publishers, 1973.

Shivers, Frank R., Jr. *Bolton Hill: Baltimore Classic.* Baltimore: Equitable Trust Co., 1978.

Simmons, R. C. *The American Colonies from Settlement to Independence.* New York: David MacKay, 1976.

Somerset Iris and Messenger of Truth 1 (July 15, 1828): 5.

Speer, Carl, Jr. "Sanitary Engineering Aspects of Shellfish Pollution." Baltimore: The Johns Hopkins University, 1936. Master's thesis.

Starkey, Marion L. *Land Where Our Fathers Died: The Settling of the Eastern Shores.* New York: Doubleday, 1962.

State of Maryland. "Proceedings of the Governor's Conference on Chesapeake Bay." Annapolis: Westinghouse Ocean Research and Engineering Center, 1968. Typescript.

"State of the Chesapeake Bay Symposium." *University of Richmond Law Review* 29, no. 3 (May 1995), 493–788.

State Planning Commission. *Conservation Problems in Maryland.* Baltimore, 1935.

Stranahan, Susan Q. *Susquehanna, River of Dreams.* Baltimore: The Johns Hopkins University Press, 1993.

Sullivan, J. Kevin. *A Summary of the Chesapeake Bay Critical Area Commission's Criteria and Program Development Activities 1984–1988.* Annapolis: Chesapeake Bay Critical Area Commission, 1989.

Summerson, Sir John. *The Pelican History of Art: Architecture in Britain, 1530–1830.* Hamondsworth, Middlesex: Penguin, 1970.

"Symposium Issue: The Chesapeake Bay." *Virginia Journal of Natural Resources Law* 4, no. 2 (Spring 1985), 185–395.

Tate, Thad W., and David L. Ammerman, eds. *The Chesapeake in the Seventeenth Century.* Chapel Hill: University of North Carolina Press, 1979.

Tilp, Fay, and Frederick Tilp. *Chesapeake: Fact, Fiction & Fun.* Bowie, Md.: Heritage Books, 1988.

Tilp, Frederick. *This Was Potomac River.* Alexandria, Va.: Tilp, 1978.

Townsend, Richard. Diary (1851–79). Baltimore: Enoch Pratt Free Library, 1937. Typescript.

U.S. Army Corps of Engineers, Baltimore District. *Chesapeake Bay: Existing Conditions Report.* 7 vols. Baltimore, 1973.

———. *Chesapeake Bay: Future Conditions Report*, 12 vols., 1977.

U.S. Department of the Interior, Federal Water Pollution Control Administration. *Chesapeake Bay: Susquehanna River Basins Project for Water Supply and Water Quality Management.* Washington, D.C.: Government Printing Office, 1965 (revised 1966).

———. *National Estuarine Pollution Study.* Vol. 3. Washington, D.C.: Government Printing Office, 1969.

U.S. Department of the Interior, Fish and Wildlife Service. *National Estuary Study.* Vol. 3. Washington, D.C.: Government Printing Office, 1970.

Van Bibber, W. C. "The Drinking Waters in Maryland Considered with Reference to the Health of the Inhabitants." Baltimore: Medical and Chiurgical Society of Maryland, 1882. Address.

Virginia Marine Resources Commission (formerly Commissioners of Fisheries). *Annual Report.* Fiscal Years 1970; 1971; and 1972.

Virginia State Senate. *Pollution: Report of the Committee Appointed by the Governor.* Richmond: Division of Purchase and Printing, 1934.

Voigt, William, Jr. *The Susquehanna Compact, Guardian of the River's Future.* New Brunswick, N.J.: Rutgers University Press, 1972.

Wallace, McHarg, Roberts, and Todd, Inc. *Maryland Chesapeake Bay Study.* Philadelphia: Maryland Department of State Planning and the Chesapeake Bay Interagency Planning Committee, 1972.

Walsh, Richard, and William Lloyd Fox. *Maryland: A History, 1632–1864.* Baltimore: Maryland Historical Society, 1974.

Warner, William W. *Beautiful Swimmers: Watermen, Crabs and the Chesapeake Bay.* Boston: Little, Brown, 1976.

Warren, Marion E. *Bringing Back the Bay: The Chesapeake in the Photographs of Marion E. Warren and the Voices of its People.* Baltimore: The Johns Hopkins University Press, 1994.

Water Resources Policy Committee. *A Water Policy for the American People.* Vol. I. Washington, D.C.: Government Printing Office, 1950.

Watson, Mark S. "The Chesapeake Country's Life Revives." *The Baltimore Sun,* July 13, 1930.

Wennersten, John R. *The Oyster Wars of Chesapeake Bay.* Centreville, Md.: Tidewater Publishers, 1981.

Wertenbaker, Thomas J. *The Golden Age of Colonial Culture,* 2d ed. Ithaca, N.Y.: Cornell University Press, 1959.

———. *Norfolk: Historic Southern Port.* 2d ed. Durham, N.C.: Duke University Press, 1962.

———. *The Shaping of Colonial Virginia.* New York: Russell and Russell, 1958.

White, Christopher P. *Chesapeake Bay: Nature of the Estuary, A Field Guide.* Centreville, Md.: Tidewater Publishers, 1989.

White, Dan. *Crosscurrents in Quiet Water: Portraits of the Chesapeake.* Dallas, Texas: Taylor Publishing Co., 1987.

Wilstach, Paul. *Tidewater Maryland.* Cambridge, Md.: Tidewater Publishers, 1969.

Wolman, Abel. *Water, Health and Society.* Bloomington: Indiana University Press, 1969.

Wolman, Abel, and John C. Geyer. *Report on Sanitary Sewers and Waste Water Disposal in the Washington Metropolitan Region.* Baltimore: The Johns Hopkins University Press, 1962.

Wolman, Abel, John C. Geyer, and E. E. Pratt. *A Clean Potomac River in the Washington Metropolitan Area.* Washington, D.C.: Interstate Commission on the Potomac River Basin, 1957.

Index

Page numbers in italic type refer to figures or captions.

acid waste disposal, 49, 126, 127–8
Agnes, tropical storm, 171, 177–8
Agreement of Federal Agencies on Ecosystem Management in the Chesapeake Bay, 216
air pollution, 180–2
algae: blooms caused by nutrients in Bay, 112, 177–8, 179; concentrations of in Bay, 186; and sunlight, freshets, and water temperature, 186
Allen, Gov. George, 184
Allied Chemical Corp., 172–7
Alsop, George, 50
American Revolution, 31
Anacostia River, toxic pollution of, 183
anadromous fish, 162, 192
Angle, Prof. J. Scott, 186
animal waste storage facilities, 212–3
Annapolis: founding, 26; layout, 49
Arnick, Del. John S., 144
Atomic Energy Commission, 151–2
automobiles, air pollution caused by, 180, 183–4

Back Basin, Baltimore, 34
Back River, 47, 70, 89, 112, 124–5
Back River sewage treatment plant, 47, 70–1, *88*, 88–9, 125, 221
bacterial pollution, 3, 162–3
Baker, Joel, 181
Baker, William C.: on Bay cleanup efforts, 219; on Bay crab fishery, 194; on EPA Chesapeake Bay study, 180
Baliles, Gerald L.: on Bay cleanup efforts, 229; on Chesapeake Bay Agreement (1983), 200–1; on Chesapeake Bay Agreement (1987), 210–1; on value and use of Bay, 4–5
ballast discharging, 56, 57, 61
Baltimore: canning in, 32–3; "coffee fleet," 32; founding, 25; growth in nineteenth century, 30–3; health police, 42; lazaretto, 42; origin of name,

31; population, 26, 31, 32, 35, 47; public water supply, 34–5, 36; sewage system, 46–7, 87–90, 221; sewage treatment, 84–90; trade volume after Civil War, 58; waste disposal, 47–9
Baltimore, Battle of (War of 1812), 31–2
Baltimore, Lord, 24
Baltimore clippers, 31, 32
Baltimore Gas & Electric Co., 147, *152*
Baltimore Harbor channel, 137, 144, 145
Baltimore Harbor channel dredging. *See* channel dredging: Baltimore Harbor
Baltimore and Ohio Railroad, 32
Baltimore Public Health Department, 45
Baltimore Sewerage Commissions, 84–90, 125, 140, 162
Baltimore Water Commission, 36
Baltimore Water Company, 35
Barnes, Michael D., 218
bathing beaches, 69–70, *73*
Bay bottom grasses. *See* submerged aquatic vegetation (SAV)
Bay Declarations, 198
"Bernie Fowler Day," 195–6
Bernie Fowler's "Sneaker Index," 195, *196*, 219
best management practices for farms, 211–3
Biggins, J. C., 101
biological nutrient removal technology, 214
Bi-State Conference on the Chesapeake Bay (1977), 179–80
Bladensburg, Md., 58
blue crabs. *See* crabs
Blue Plains sewage treatment plant, 184–5, *214*, 214–5
boards of health, 42, 224
boats, pollution by, 126–7, 180–1
Bond, Dr. Thomas E., 45–6
boundary between Maryland and Virginia: in Chesapeake Bay, *13*, 29, 81; in Potomac River, 29, 51, 81